Environmental and Quality Systems Integration

William C. Culley

Lewis Publishers

Boca Raton Boston London New York Washington, D.C.

Acquiring Editor: Kenneth McCombs
Project Editor: Albert W. Starkweather, Jr.
Cover design: Denise Craig

Library of Congress Cataloging-in-Publication Data

Culley, William C.
   Environmental and quality systems integration / William C. Culley
       p. cm.
   Includes bibliographical references and index.
   ISBN 0-56670-288-7 (alk. paper)
   1, ISO 14000 Series Standards. 2. ISO 9000 Series Standards. 3. Production mangements I. Title.
T155.7.C85 1998
658.4′08—dc21

                    98-12630
                    CIP

Author

William C. "Chuck" Culley is an Environmental, Health, and Safety Manager with the Electronics OEM Division of Raychem Corp., headquartered in Menlo Park, CA.

He holds a Bachelor of Science Degree in Organic Chemistry from the University of California at Los Angeles and certification in Hazardous Materials Management from the University of California at Santa Cruz. He is a Certified Safety Professional (CSP), a Certified Hazardous Materials Manager (CHMM), and a member of the American Society of Safety Engineers (ASSE).

Mr. Culley guided Raychem's PolySwitch Product Group to an ISO 14001 certification in early 1997. His successful implementation of the new Environmental Management Standards came as a result of utilizing the existing ISO 9001 Quality Management System framework already implemented at PolySwitch. Through a careful analysis and comparison of the two standards, he was able to merge and integrate a vast majority of the ISO 14001 documentation requirements into the ISO 9001 framework. His efforts in this area have made him an advocate of the complete integration of his company's quality and environmental, health and safety programs on a worldwide basis.

His career interests include the implementation of Business Recovery Programs, the merging of Product Stewardship and Design Review Programs, and pollution prevention activities in terms of air emissions and solid waste reduction and elimination.

Acknowledgments

Foremost, I would like to thank my wife, Glenda, for her love and support during the many late nights spent writing this book, and her encouragement in pursuing this opportunity. I also would like to extend my appreciation to the following people: Jerry Jones, Raychem Corp.'s EH&S Director, for introducing ISO 14001 to me and providing support during my program implementation and development; Jeff Braggin, Raychem QA Manager, for his support and recommendations for integrating ISO 14001 with ISO 9001; Peter Brooks, my former manager, for his strong commitment to sustainable development and his encouragement to me to do "what is right" for the environment; and Gary Cottrell and Dr. Anne-Marie Warris, LRQA Environmental Auditors, for the valuable auditing experience I learned from them in pursuing ISO 14001 certification.

Preface

For several years, Environmental, Health, and Safety (EH&S) professionals have been attempting to penetrate the "Quality Market." Although progress for many in the field has been slow up until the last two years or so, help is on the way. With final issuance of the ISO 14000 standards in the fall of 1996, EH&S Professionals have found a new door opening into their company's quality system. The major obstacle, however, will be determining how EH&S programs can be integrated into the existing quality programs such that it will not place a major burden on the existing management and document systems. The fear that another mountain of documentation will result from implementing ISO 14001 is never off a senior manager's mind. Because of this fear and concern, many companies have not and will not choose to integrate the two systems. This book is intended to provide a guideline into how the integration process can occur.

Introduction

Many EH&S professionals have attempted to show senior management that their work and efforts can help a company improve its overall processes and, thus, increase profitability. Many businesses are now seeing that good environmental performance can be an extra business opportunity and not an extra business overhead. The EH&S Manager needs to become a salesman and convince senior management that there is a "hard dollars and cents" Return on Investment and that the investment of time, personnel, and finances is well worth it.

Several firms and the International Standardization Organization (ISO) have recommended integrating ISO 14001 and ISO 9000, but many are not sure how to effectively go about starting such a project. This book is intended to provide you with a practical *"how to"* method for integrating your Environmental and Quality Management Systems. The overall intent is to take an operation with multiple management systems, blend them, and end up with one operational or management system. This "blending" or integration process is essential for streamlining processes, improving profits, and gaining an advantage in today's global markets. This is all part of what is now commonly known in business circles as *"strategic standardization."*

As you go through this process, it is highly recommended that your time and effort also include the integration of your occupational health and safety programs, especially if your responsibilities include them along with environmental programs. With the potential for international occupational health and safety standards looming in the near future, it might make sense to gain a head start. Many large organizations, however, will have a much more difficult time doing this because of the organizational separation of the environmental departments from the occupational safety and health departments. If this is the case, there still would be added benefits for each department to integrate their respective systems and to begin working together for complete merger over a longer period of time.

Working through this process will have some side-benefits. The first is obvious — preparation for an ISO 14001 EMS audit. As we deal with each ISO element in this book, a basic question will be asked periodically:

What will an ISO Accredited Auditor look for?

Second, as you work through this process of integration, management will notice the following overall benefit:

Quality Systems + EH&S Systems = Business Excellence

A third benefit is in the area of international trade which will be discussed in further depth in a later chapter.

Another important factor in the integration process involves the development of a working relationship between the EH&S Professional and the Quality Assurance Manager. It is vital that these two individuals (or groups) work together to accomplish the integration of their respective management systems. The EH&S Professional must not only be knowledgeable of his company's quality system, but also tactful and "sensitive" to the management structure which has been built up by the Quality Assurance Manager. The EH&S Manager must view the QA Manager as a mentor, consultant, advisor, and professional resource for learning the nuances of the ISO management system. The QA Manager can provide overall guidance by assessing many of the management systems such as document control, records maintenance, auditing, management reviews, corrective and preventive actions, and nonconformances and be able to do this without additional training. The potential for successfully completing an ISO 14001 assessment with minimal need for improvement will be greatly enhanced.

By comparing the various elements of ISO 9001 and ISO 14001, as outlined in Appendix B, both the QA Manager and EH&S Professional will gain a better understanding of the tremendous potential for integrating their two management systems; thus, this book is intended to be useful for both professions. The successful completion of an integration will require the EH&S Professional and the QA Manager to develop new skills and an understanding of their counterpart's professional arena, but it is most important for the EH&S Professional to become familiar with the QA system in order to have the knowledge and skill to conduct an assessment. The importance for the QA Manager to gain an understanding of the environmental arena is not as critical since, again, we are mainly concerned about a management system and not a legal compliance audit. However, since this book is also intended to be used by a QA Manager, some introductory material on some basic environmental systems is presented in Chapter 2.

This book will focus primarily on ISO 14001, the specification standard, and its accompanying guidance document (ISO 14004) with very minimal discussion of the possible implications surrounding the other supporting ISO Standards — Life Cycle Analysis, Labeling, Environmental Performance Evaluation, and Product Standards. Although the primary subject of this book is the integration of ISO 14001 with ISO 9000, the consideration to integrate other Environmental Management Systems such as BS 7750, Eco-Management Audit Scheme (EMAS), and your own particular business' program into other quality systems should not be ignored. As we evaluate some of these quality systems, you may see where environmental management can play a significant role in making business decisions; in fact, many quality system assessments are now being modified to include environmental, health, and safety issues as businesses continue to gain an understanding of the importance Environmental, Health, and Safety can play in managing a quality program.

Since the basic philosophy behind this book is the development of a single operational/management system, it would be logical to first spend time discussing and providing a background on the quality and environmental "movements" and how various parts of the world have viewed both systems not only from an industry standpoint, but also nationally. Figure 1 demonstrates the intent of the book — blending environmental and quality programs into a single management system.

The focus will be on those primary quality and environmental management systems currently in use around the world which had the strongest influence on the

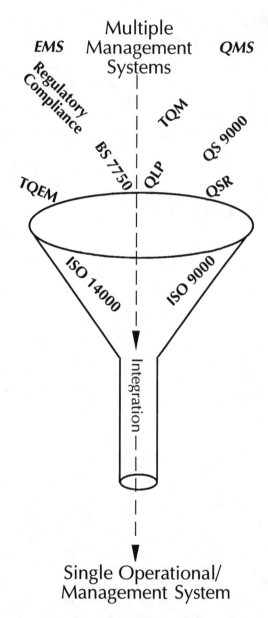

FIGURE 1. Blending quality and environmental management into a single operational/management system.

development of both ISO 9000 and ISO 14000. It also is important to have this background if your company is currently conducting or is planning to conduct trade on an international scale. Integrating environmental and quality systems is much like a marriage — the more you understand each other's background and "family" history, the better able you are to make the marriage work and develop as one.

Contents

Part III Planning

Part IV Implementation and Operation

Part V   Checking and Corrective Action

Part VI   Management Review

Part VII   Conclusion

Tables

Figures and Charts

FIGURES

CHARTS

Part I

Background on QMS and EMS Programs

1 The Quality Movement

1.1 INTRODUCTION

Over the past 20 years or so quality systems have become a major focus in business circles. A business' customer base is continually demanding better product at a lower cost and, in doing so, has forced executive management to rethink how they operate their business from top to bottom. Many of the larger firms such as those in Table 1.1 have had quality auditing programs for their suppliers that have resulted in a steady stream of auditors visiting a business site. This "revolving door" of auditors has given a large number of firms a much stronger foundation for developing their own quality system. It has also given them a "jump start" on implementing ISO 9000 and some of the more stringent quality programs required by customers.

Apart from a few of the major corporations, a structured quality system early on was at best primitive and management was somewhat shortsighted to see the long-term benefits. This short-term thinking has caused the United States to fall behind the Europeans and Japanese in the quality of its manufacturing and in implementing ISO 9000. It has only been recently that U.S. firms have begun to regain the competitive edge. Unfortunately, this "wait and see" approach to new ideas, standards, and methods of doing business has been the norm for U.S. companies. It continues to dominate some of the thinking with the newly released ISO 14000 Environmental Management Standards.

This chapter is intended primarily for the environmental manager. It will focus on giving an overview of the various quality programs in place on a global basis, discuss various national programs, and provide a basic understanding of how the "quality movement" evolved.

1.2 THE UNITED STATES

1.2.1 AMERICAN NATIONAL STANDARDS INSTITUTE (ANSI)

ANSI is a private federation founded in 1918 that consists of manufacturing and service-oriented businesses, professional societies, government agencies, and consumer and labor groups serving the private and public sectors. One of its functions is to coordinate the United States' involvement in the standards systems. ANSI is the U.S. representative to the International Standardization Organization (ISO) which has developed both the ISO 9000 and ISO 14000 Standards. The well-known "ANSI Standards" have provided conformity in many areas, primarily in the area of health and safety. One of these publications, done jointly with ASQC, is a first attempt at integrating quality and environmental management systems. Known as ANSI/ASQC

TABLE 1.1
Examples of Corporate Quality Systems

Industry	Quality System
Motorola	Quality Systems Review (QSR)
AT&T	Quality Leadership Program (QLP)
U.S. Automotive	Quality System Requirements (QS-9000)
ASMO	Supplier Quality Audit (SQA)
Xerox	Quality Improvement Process (QIP)

TABLE 1.2
Basic Contents of ANSI/ASQC E4–1994

(1) General Provisions Evaluation of Environmental Data	(3) Part B: Collection and Evaluation of Enivronmental Data
1.1 Introduction	3.0 General
1.2 Purpose and Content	3.1 Planning and Scoping
1.3 Scope and Field of Application	3.2 Design of Data Collection Operations
1.4 Normative References	3.3 Implementation of Planned Operations
1.5 Definitions	3.4 Assessment and Response
	3.5 Assessment and Verification of Data Usability
(2) Part A: Management Systems	**(4) Part C: Design, Construction, and Operation of Environmental Technology**
2.0 General	4.0 General
2.1 Management and Organization	4.1 Planning
2.2 Quality System and Description	4.2 Design of Systems
2.3 Personnel Qualification and Training	4.3 Construction/Fabrication of Systems and Components
2.4 Procurement of Items and Services	4.4 Operation of Systems
2.5 Documents and Records	4.5 Assessment and Response
2.6 Computer Hardware and Software	4.6 Verification and Acceptance of Systems
2.7 Planning	
2.8 Implementation of Work Processes	
2.9 Assessment and Response	
2.10 Quality Improvement	

Printed with the permission of the American Society of Quality Control, 611 W. Wisconsin Ave., Milwaukee, WI 53201–3005. Copyright *ASQC Quality Press.*

E4–1994, *Specifications and Guidelines for Quality Systems for Environmental Data Collection and Environmental Technology Programs,* its purpose is to provide a system for standardizing a quality management system for environmental data collection and evaluation, and environmental technology design, construction, and operation. Table 1.2 describes the contents of ANSI/ASQC E4–1994.

1.2.2 AMERICAN SOCIETY OF QUALITY CONTROL (ASQC)

Founded on Feb. 16, 1946, ASQC's mission is to facilitate quality improvement tools. This includes providing programs to share information on statistical process control, quality cost measurement and control, total quality management (TQM), failure analysis, and zero defects. ASQC administers, on behalf of ANSI, the U.S. TAG to ISO/TC 176 and TC 207 and, in conjunction with ANSI, is the primary source of ISO 9001 and ISO 14001 information and documentation. Where many industry quality programs have provided a defining structure to the quality system, it has been the ASQC quality improvement tools that have actually helped a business to complete the process.

1.2.3 TOTAL QUALITY MANAGEMENT

As a pre-ISO 9000 system, TQM is a uniquely American view on how to manage quality and, in its early stages, was the United States' response to Total Quality Control (TQC) begun in Japan. TQM, however, was about 20 years behind TQC (late 1980s vs. early 1960s) in its development. Even after almost 10 years it has still failed to reach the level of maturity TQC has attained, and most likely never will. This is primarily due to three factors: first, the early, short term thinking and "get rich quick" philosophy of American business vs. the long term thinking of Japanese business; second, TQM's focus on management techniques whereby each manager and each employee applies their knowledge and skills to assure high work productivity, then high product quality, and, ultimately, high customer satisfaction (whereas TQC has developed the integration of product and process development through engineering efforts); and, third, the introduction of ISO 9000 has provided a much more defined structure for the establishment of a quality management system. As more structured programs become established, it can be expected that TQM as a system in itself will steadily lose ground and potentially disappear as a viable quality management system.

The TQM philosophy was born in the early 1980s as a result of the frantic search by U.S. industry for a system that would steer it in a direction that would correct its short-term thinking and narrow the gap between it and Japanese industry. TQM, as a result, has "fathered" various quality assurance programs now in use in U.S. industry, such as those in Table 1.1. It has also had its influence on the development of other management systems, one of which is Total Quality *Environmental* Management (see Chapter 2).

1.2.4 QS-9000

Quality Systems Requirements, QS-9000, was developed by the "Big Three" of the U.S. automotive industry — General Motors, Ford, and Chrysler. The purpose of the program is to harmonize and standardize the product quality of their suppliers through continuous improvement as it impacts product safety (preventing defects) and cost reduction (reducing variations and minimizing waste). What QS-9000 does is take ISO 9000 to a higher level of requirements that are more sector (e.g., automotive) and customer-specific as compared to ISO 9000's general international requirements.

One important feature of QS-9000 is its inclusion of some environmental requirements not found in ISO 9000, but are found in ISO 14001. The U.S. automotive industry has recognized the importance of including environmental management into quality management. These environmental requirements are found in QS-9000, Section 2, *Sector-Specific Requirements*, and Section 3, *Customer-Specific Requirements*. Along with ISO 9000 and ISO 14001, QS-9000 will be one of the primary management systems evaluated in this book in terms of quality and environmental integration.

1.2.5   MALCOLM BALDRIDGE NATIONAL QUALITY AWARD

In August 1987, President Ronald Reagan signed Public Law 100–107, the Malcolm Baldridge National Quality Award. The purpose of the award was threefold: (1) to promote the awareness of quality excellence; (2) to recognize quality achievements of U.S. businesses; and (3) to publicize successful quality strategies. The responsibility for administering the award was bestowed on the National Institute of Standards and Technology (NIST).

The quality criteria listed in Table 1.3 can assist in the development of an environmental management system. A strong commitment to environmental management has been beneficial to several companies in their pursuit of this award. Specifically, two past winners of the National Quality Award indicated their environmental management systems greatly influenced and enhanced their overall quality systems.

Although the Baldridge Quality Award has its primary focus on quality, it also enhanced their environmental programs. Xerox, a 1989 winner, began formalizing its commitment to the environment when it established a corporate environmental, health, and safety department. This environmental commitment has been greatly reinforced by the company's Leadership Through Quality program, launched in 1984, which led to Xerox Business Products and Systems winning the National Quality Award. Motorola, a 1988 winner, began implementing an environmental information system as a tool in its overall Six Sigma quality systems. The Environmental Information Management System (EIMS) has provided a mechanism for measuring environmental performance and the development of an associated sigma metric for the continual improvement of Motorola's environmental programs.[1]

1.3   THE EUROPEAN UNION

Since it might be expected that most of the readers of this book will be from the United States, it might be important for you to gain a brief understanding of how the European nations operate in their implementation of quality and environmental management systems. This is important because of the primary leadership role the European Union has assumed in the development of the international standards and the potential impact this may play as trade negotiations between the United States and Europe rise to a new level of understanding.

As the major player in the endorsement and establishment of international standards, the European Union consists of 15 European nations much the same way as the United States consists of 50 individual states. These 15 member nations are:

TABLE 1.3
The Malcolm Baldrige National Quality Award 1997 Criteria

1997 Categories/Items	Point Values
1.0 Leadership	**110**
1.1 Leadership System	80
1.2 Company Responsibility and Citizenship	30
2.0 Strategic Planning	**80**
2.1 Strategy Development Process	40
2.2 Strategy Deployment	40
3.0 Customer and Market Focus	**80**
3.1 Customer and Market Knowledge	40
3.2 Customer Satisfaction and Relationship Enhancement	40
4.0 Information and Analysis	**80**
4.1 Selection and Use of Information and Data	25
4.2 Selection and Use of Comparative Information and Data	15
4.3 Analysis and Review of Company Performance	40
5.0 Human Resource Development and Management	**100**
5.1 Work Systems	40
5.2 Employee Education, Training, and Development	30
5.3 Employee Well-Being and Satisfaction	30
6.0 Process Management	**100**
6.1 Management of Product and Service Processes	60
6.2 Management of Support Processes	20
6.3 Management of Supplier and Partnering Processes	20
7.0 Business Results	**450**
7.1 Customer Satisfaction Results	130
7.2 Financial and Market Results	130
7.3 Human Resources Results	35
7.4 Supplier and Partner Results	25
7.5 Company-Specific Results	130
Total Points:	**1000**

Germany, France, the United Kingdom, Italy, Spain, Belgium, The Netherlands, Sweden, Denmark, Norway, Finland, Switzerland, Italy, Austria, and Liechtenstein. They have joined together and recognize the authority of the European Union just as the U.S. states recognize the authority of the United States Federal government.

However, not all of the European Union member nations (e.g., the United Kingdom) fully endorse the extension of the European Union as a strong federal government structure and do not desire to give up their own identity. Table 1.4 shows the comparative rulemaking bodies for the European Union compared to the United States.

TABLE 1.4

Comparative U.S. and European Rulemaking Bodies

United States	European Union
Executive Branch	European Commission (EC)
Senate	European Council
House of Representatives	European Parliament

1.3.1 EUROPEAN COMMISSION (EC)

This Commission's primary responsibility is to propose legislation that will impact the environment of the member nations. Legislation is normally issued as a regulation or a directive.

Regulations are laws that are binding in all member nations without the implementation of any national legislation. An example of a regulation is: EC 1836/93, *EMAS, Eco-Management Audit Scheme. Directives* are also laws, but each member can implement the requirements in any manner that will achieve the desired end effect that may include the implementation of national legislation.

1.3.2 BRITISH STANDARDS INSTITUTE

The British Standards Institute (BSI) was founded in 1901 to assist in the development of British national standards and was the very first national standards body in the world — predating ANSI, ASQC, and ISO. BSI participates as an active member of ISO, CEN (Comitee Europeen de Normalization or Committee for European Standardization), and CENELEC (Comitee Europeen de Normalization Electrotechnique or European Committee for Electrotechnical Standardization). It provides guidance on the development of the European standards as the European Union marches towards a single market.

As mentioned earlier, the United Kingdom does not completely endorse the growing influence of the European Union. It is expected that the British Standards Institute will continue to attempt to take a leadership role in the issuance of standards for application worldwide.

1.4 CANADA

First established in 1919, the Canadian Standards Association (CSA) is Canada's primary standards developer. It certifies and provides registration through its Quality Management Institute (QMI) division. The Canadian Standards Association has been one of the principal players in the development of international standards, such as ISO 9000 and ISO 14000 (primarily on the development of ISO 14040, *Life Cycle Assessment*). As the Canadian counterpart of ANSI, it also supports the International Electrotechnical Commission (IEC) and has been a primary driver of standards in the areas of design and safety. The famous CSA logo is carried on millions of products throughout the world.

1.5   JAPAN

Many people would argue that the global "quality movement," as we know it, began in Japan after World War II when W.S. Magil of Bell Labs introduced Statistical Quality Control (SQC) to a Japanese industry undergoing rebuilding. However, throughout most of the 1950s and early 1960s, product labeled as "Made in Japan" was considered to be cheap and of inferior quality. Since the mid 1960s, however, that has changed 180 degrees; a quality manufactured product is not only a German hallmark, but now a Japanese one as well. The Japanese not only produce quality, but also demand high quality from their suppliers.

As mentioned earlier when discussing TQM, Total Quality Control originated primarily in Japan in the early 1950s and was developed over the years by utilizing the quality teachings of Deming and others. The Deming Prize originated in Japan and is awarded by the Japan Union of Scientists and Engineers (JUSE). TQC has been the primary instrument for implementing "Kaizen" (continuing improvement), Ishikawa (cause and effect), and the "Taguchi methods" (design of experiments).

1.6   INTERNATIONAL STANDARDIZATION ORGANIZATION

Established in 1947, the International Standardization Organization (ISO) is a non-governmental, worldwide federation consisting of some 90 national bodies. It promotes the development of standardization and related activities with the view to facilitating international exchange of goods and services in most standardization fields. The exceptions are electrical and electronic engineering that are the responsibility of the International Electrotechnical Commission (IEC). The ISO accomplishes this by:

- improving product quality and reliability.
- better compatibility and operability of products and services.
- improving environmental, health, and safety protection.
- simplification for improved usability.
- reducing the number of models and thus cost.
- increasing distribution efficiency.

In 1979, ISO formed Technical Committee (TC) 176 to harmonize the preponderance of national and international standards in the quality field. The result was the release in 1987 of ISO 9000, *Quality Management and Quality Assurance.* Since 1987, certification to the ISO 9000 series of standards has been growing in momentum and has become, in many instances, a requirement to conduct business. It has had its sources of controversy; many businesses have resisted ISO 9000 certification because of the initial and upkeep expenses and the overall time commitment involved.

TABLE 1.5
The ISO 9000 Series of Standards

Description	Standard	Primary Focus
Quality Management and Quality Assurance Standards	ISO 9000	Guidelines for Selection and Use
Quality Systems — Model for Quality Assurance in Design, Development, Production, Installation, and Servicing	ISO 9001	Design, Manufacturing, Installation, and Service
Quality Systems — Model for Production, Installation, and Servicing	ISO 9002	Production and Installation
Quality Systems — Model for Final Inspection and Test	ISO 9003	Final Product Inspection and Test
Quality Management and Quality Assurance Elements	ISO 9004	Guidelines

Additionally, as Table 1.1 points out, many firms have had their own world-class quality programs in place and, therefore, feel they have nothing to gain from implementing an ISO 9000 certification process. ISO 9000 is a set of five related standards tailored for the user and dependent on the scope or type of business. Table 1.5 gives a brief description of the overall system.

Being ISO 9000 certified has not necessarily resulted in fewer customer quality audits. However, as time has passed, the resistance to ISO 9000 certification has dropped off sharply. As acceptance of the system has gained momentum, many firms have begun to recognize certification to the international standards as the only way to unlock new marketing areas despite the high quality systems that they may already have in place.

1.7   THE DEMING PRIZE

For more than 40 years of "preaching" quality, W. Edwards Deming focused on the need to empower the average worker to take control of the quality of his work and for management to accept the blame for everything that goes wrong in business. Although Deming had been "preaching" management responsibility, it was not until 1980 that he actually became known to U.S. industry. For the previous thirty years, he worked primarily with Japanese industry to be more effective in utilizing the techniques of statistical quality control (SQC) introduced into Japan just after World War II by W. S. Magil.

Since 1951, the "Deming Application Prize" has been given by the Japan Union of Scientists and Engineers (JUSE) to companies who display outstanding quality programs. The Deming Prize is given to outstanding individuals in the quality field. The award is an international award and is only given once a year. Deming's 14 Points for Management, initially developed in the 1960s, is basically an abbreviation of his "preaching."

1.7.1 DEMING'S 14 POINTS FOR MANAGEMENT*

1. Create constancy of purpose toward improvement of product and service, with the aim to become competitive and to stay in business, and to provide jobs.
2. Adopt the new philosophy. We are in a new economic age. Western management must awaken to the challenge, must learn their responsibilities, and take on leadership for change.
3. Cease dependence on inspection to achieve quality. Eliminate the need for inspection on a mass basis by building quality into the product in the first place.
4. End the practice of awarding business on the basis of price tag. Instead, minimize total cost. Move toward a single supplier for any one item, on a long-term relationship of loyalty and trust.
5. Improve constantly and forever the system of production and service, to improve quality and productivity, and thus constantly decrease costs.
6. Institute training on the job.
7. Institute leadership. The aim of supervision should be to help people and machines and gadgets to do a better job. Supervision of management is in need of overhaul as well as supervision of production workers.
8. Drive out fear, so that everyone may work effectively for the company.
9. Break down barriers between departments. People in research, design, sales, and production must work as a team, to foresee problems of production and in use that may be encountered with the product or service.
10. Eliminate slogans, exhortations, and targets for the work force asking for zero defects and new levels of productivity. Such exhortations only create adversarial relationships, as the bulk of the causes of low quality and low productivity belong to the system and thus lie beyond the power of the work force.
11. a. Eliminate work standards (quotas) on the factory floor. Substitute leadership.
    b. Eliminate management by objective. Eliminate management by numbers, numerical goals. Substitute leadership.
12. a. Remove barriers that rob the hourly worker of his right to pride of workmanship. The responsibility of supervisors must be changed from sheer numbers to quality.
    b. Remove barriers that rob people in management and in engineering of their right to pride of workmanship. This means, *inter alia*, abolishment of the annual merit rating and of management by objective.
13. Institute a vigorous program of education and self-improvement.
14. Put everybody in the company to work to accomplish the transformation. The transformation is everybody's job.

1.8 CONCLUSION

There is a tremendous amount of diversity in the quality world. As expected, the more industrialized nations have very advanced programs and have organizations whose primary purpose is to develop and promote quality standards on a national and international scale. A large percentage of small to medium-sized businesses have quality management systems in place and/or are aware of the need of implementing such a program. In contrast, quality programs in less industrialized nations are primarily focused on quality management systems established by a few of their prominent national industries. Small to medium-sized business either do not have the resources and even the awareness of the various quality programs in existence and the need for one.

An environmental manager must become aware of the quality management systems his firm is using or pursuing. It is not expected that an environmental manager should gain an exhaustive understanding of his firm's quality program requirements. The more you understand, the easier it will be to integrate environmental decisions into the quality decisions.

The environmental manager may also require knowledge of several quality systems if the firm is multinational in its operations. The same will be true for a quality manager's need to understand the environmental arena — this will be discussed in the next chapter.

2 The Environmental Movement

2.1 INTRODUCTION

Man's concern over the environment goes back several hundred years, but has been most noticeable since the mid-19th century. With the coming of the Industrial Revolution, the public outcries over smoke pouring out of stacks from coal-fueled factories, and, ultimately, the expansion of the petroleum industry, the "Environmental Movement" actually began to take hold long before the "Quality Movement." This chapter is primarily intended for the quality manager who may have little knowledge of the environmental arena. We will discuss some of the primary global environmental management systems (legislated or voluntary standards) that currently provide most of the impetus to the environmental management systems movement and the development of ISO 14001, as well as the current status of various national environmental programs throughout the world.

2.2 THE UNITED STATES

2.2.1 AMERICAN PETROLEUM INSTITUTE

Years before widespread environmental legislation was passed in the U.S., the petroleum industry had already recognized the importance of its role in maintaining the environment. In 1919 the American Petroleum Institute (API) was founded as a result of an awareness by the petroleum industry that a lack of standardization had been a primary contributor to the shortages experienced by industry around the turn of the century. Consisting of the petroleum and allied industries, the API has a four-fold mission, one of which is the enhancement of the environmental, health, and safety performance of the petroleum industry. With additional pressure on the oil industry coming from regulatory agencies, the API began to invest heavily in environmental stewardship. The result was in 1990 the creation of STEP — *Strategies for Today's Environmental Partnership*. STEP has become the framework for the petroleum industry to improve its environmental, health, and safety performance. This framework is built on API's Environmental, Health, and Safety Mission and Guiding Principles. It is a basic system for preventing pollution, conserving natural resources, measuring progress, promoting product stewardship, maintaining crisis readiness, addressing community concerns, working with government agencies, and reporting results to develop responsible rules to protect the environment, the public, and employees.[2] All of these principles are very similar to the ISO 14001 framework and, according to API management, have no inconsistencies with ISO 14001. Petroleum-based industries who are implementing STEP already have most of the

ISO 14001 elements in place. With this foundation in place, API is also encouraging the petroleum industry constituents (gasoline stations, refineries, drilling operations, etc.) to become registered to ISO 14001. Implementing ISO 14001 would be most advantageous to those conducting or considering conducting international business and wish to be recognized for conformance to an international environmental management system. Additionally, you will also notice the API principles to be very much a policy statement which incorporates quality principles as well.

2.2.1.1   American Petroleum Institute Environmental, Health and Safety Mission and Guiding Principles*

The members of the American Petroleum Institute are dedicated to continuous efforts to improve the compatibility of our operations with the environment while economically developing energy resources and supplying high quality products and services to consumers. We recognize our responsibility to work with the public, the government, and others to develop and to use natural resources in an environmentally sound manner while protecting the health and safety of our employees and the public. To meet these responsibilities, API members pledge to manage our businesses according to the following principles using sound science to prioritize risks and to implement cost-effective management practices:

1. To recognize and to respond to community concerns about our raw materials, products, and operations.
2. To operate our plants and facilities, and to handle our raw materials and products in a manner that protects the environment, and the safety and health of our employees and the public.
3. To make safety, health and environmental considerations a priority in our planning, and our development of new products and processes.
4. To advise promptly appropriate officials, employees, customers and the public of information on significant industry-related safety, health, and environmental hazards, and to recommend protective measures.
5. To counsel customers, transporters and others in the safe use, transportation, and disposal of our raw materials, products, and waste materials.
6. To economically develop and produce natural resources and to conserve those resources by using energy efficiently.
7. To extend knowledge by conducting or supporting research on the safety, health, and environmental effects of our raw materials, products, processes, and waste materials.
8. To commit to reduce overall emissions and waste generation.
9. To work with others to resolve problems created by handling and disposal of hazardous substances from our operations.

---

* Printed with the permission of the American Petroleum Institute, 1229 L Street, NW, Washington, D.C. 20008

10. To participate with government and others in creating responsible laws, regulations and standards to safeguard the community, workplace and environment.
11. To promote these principles and practices by sharing experiences and offering assistance to others who produce, handle, use, transport or dispose of similar raw materials, petroleum products and wastes.

2.2.2 THE ENVIRONMENTAL PROTECTION AGENCY

The environmental regulatory movement in the United States actually began several decades before the creation of the EPA and started with the Federal Insecticide, Fungicide, and Rodenticide Act in 1947 and the Air Pollution Control Act of 1955. However, it was not until the late 1960s that the government began to take more aggressive and pragmatic steps to control environmental pollution. In 1969, the U.S. Federal government passed a statute known as the National Environmental Policy Act (NEPA) which was enacted in response to a growing concern over environmental harm resulting from population growth, high-density urbanization, industrial expansion, resource exploitation, and new and expanding technological advances. NEPA declares that it is the continuing policy of the federal government "to use all practicable means and measures, including financial and technical assistance, to create and maintain conditions under which man and nature can exist in productive harmony, and fulfill the social, economic, and other requirements of present and future generations of Americans."

In the following year (1970) President Richard M. Nixon signed an executive order that consolidated several federal environmental activities into a single agency. It was at this point that the Federal Environmental Protection Agency (EPA) was born and it has been nonstop since. Although the EPA lacks a statutory charter or even an overall policy, its purpose is to "protect and enhance our environment today and for future generations to the fullest extent possible under the laws enacted by Congress. The agency's mission is to control and abate pollution in the areas of air, water, solid waste, pesticides, radiation, and toxic substances. Its mandate is to mount an integrated, coordinated attack on environmental pollution in cooperation with state and local governments."

Although the primary focus of the EPA has been regulatory compliance under major statutes (Table 2.1), rather than management systems development, it has helped build the groundwork for some of the major U.S. industries' development of an environmental management system.

The Environmental Protection Agency was involved in the development of ISO 14001 and primarily engineered two of the requirements now in the standards: pollution prevention and compliance management. In its evaluation of any environmental system, the "bottom line" for the EPA includes the following questions:

- Can it be a vehicle for going beyond compliance through pollution prevention?
- Can it assist the organization in improving its compliance?
- Will other stakeholders buy into the environmental management system (e.g., ISO 14001) as an alternative to "command and control"?

TABLE 2.1

U.S. Environmental Protection Legislation

Year	Statute	Purpose
1947	Federal Insecticide, Fungicide, and Rodenticide Act	Screening of the toxic ingredients used in pesticides to ensure they do not present unnecessary hazards to human health or "nontargeted" species
1970	Clean Air Act (amendments)	Requires EPA to set uniform federal ambient air standards through emission controls on new stationary sources, hazardous air pollutants, and new motor vehicles
1972	The Federal Water Pollution Control Act (also known as the "Clean Water Act")	Into the waters of the United States: (a) control industrial discharges; (b) control and prevent spills of oil and hazardous substances; (c) regulate discharges of dredge and fill materials; and (d) provide financial assistance for construction of publicly owned sewage treatment works
1976	The Solid Waste Act (also known as the "Resource Conservation and Recovery Act")	"Cradle to grave" management (including storage, treatment, and disposal) of hazardous waste
1976	Toxic Substance Control Act	Imposes broad regulatory control over all chemicals produced or used in order to eliminate the risk of human or environmental exposure to an untested chemical
1980	The Comprehensive Environmental Response Compensation and Liability Act (also known as "Superfund")	Control inadequate authority and funding in existing environmental legislation for dealing with uncontrolled and abandoned hazardous waste sites
1986	The Emergency Planning and Community Right-to-Know Act	Requires public notification of Extremely Hazardous Substances

The EPA, however, is still not satisfied with the standard in that it is concerned ISO 14001 does not require compliance and has to date been unsuccessful in its efforts to get compliance audit requirements. It is the author's opinion that the EPA has still failed to recognize that the ISO 14001 Environmental Management Standards are voluntary for implementation worldwide and they cannot be as prescriptive from a compliance standpoint. In the United States many industries are fearful of potential compliance auditing requirements because of the legal system in the U.S. (e.g., financial and litigation issues). The result is that many firms will see this as a major obstacle to implementing a continual improvement process or any other management system such as ISO 14001. This issue will be discussed further in Chapter 3.

2.2.3 RESPONSIBLE CARE PROGRAM®

The Responsible Care Program® was established in 1988 by the Chemical Manufacturers Association (CMA) and, although it is a voluntary program, it is a requirement in order to participate in the CMA. Its members are required to:

- improve performance in health, safety, and environmental quality.
- listen and respond to public concerns.
- report their progress to the public.

The Program lists several management practices for chemical companies in the areas of pollution prevention, process safety, emergency response, employee health and safety awareness, product stewardship and general chemical management.

2.2.3.1 Guiding Principles of Responsible Care®

1. To recognize and respond to community concerns about chemicals and our operations;
2. To develop and produce chemicals that can be manufactured, transported, used, and disposed of safely;
3. To make health, safety, and environment considerations a priority in our planning for all existing and new products and processes;
4. To report promptly to officials, employees, customers and the public, information on chemical-related health or environmental hazards and to recommend protective measures;
5. To counsel customers on the safe use, transportation, and disposal of chemical products;
6. To operate our plants and facilities in a manner that protects the environment, and the health and safety of our employees and the public;
7. To extend knowledge by conducting or supporting research on the health, safety and environmental effects of our products, processes and waste chemicals;
8. To work with others to resolve problems created by past handling and disposal of hazardous substances; and
9. To participate with government and others in creating responsible laws, regulations, and standards to safeguard the community, workplace, and environment.

As you become more familiar with ISO 14001, you will notice the compatibility with Responsible Care®. A key difference is that Responsible Care® is a series of very specific management initiatives, while ISO 14001 is a very broad-based environmental management system. The result is participation in Responsible Care® will not automatically meet the requirements of ISO 14001, but does provide an excellent start. Of major concern to many members of the CMA, however, is whether or not ISO 14001 will ultimately even replace Responsible Care®. Specific key differences between Responsible Care® and ISO 14001 are in the areas of identifying environmental aspects, monitoring and measurement as it applies target progress and, thus, continual improvement, auditing, and management review.

2.3 THE EUROPEAN UNION

Various members of the European Union have been primary drivers of the ISO 14001 Environmental Management Standards. For the past two decades, many of the EU countries have had stringent environmental programs. Germany was a primary mover in the area of recycling when they passed a stringent packaging law in May 1991. The law required manufacturers to assume all responsibility for the recycling and disposal of product packaging. The rest of the European Union, however, caught on quickly. As an example, Sweden, beginning in 1994, now holds companies financially responsible for the manufacture and/or import or sale of packaging and packaged goods for the packaging materials being collected and recycled. Sweden has formed a materials company to manage the program and all companies are required to arrange for package recycling in some manner.

The EU's constantly changing regulations on the environment are continuing to make it harder and harder to conduct business on the European continent. In addition to the 200 plus EC directives, a business must be aware of the individual country's regulations and, most likely, be certified to the ISO 9000 standards. Despite the tremendous efforts to unify the members, an easing of the situation and potential standardization of environmental regulations across the EU is not expected to take place very soon.

2.3.1 ECO-MANAGEMENT AUDIT SCHEME (EMAS)

Passed in 1993 as European Commission Regulation 1836 (e.g., EC 1836/93), the Eco-Management Audit Scheme has the primary objective of promoting continuous improvement of the environmental performance of industrial activities by:

- establishing and implementing environmental policies, programs, and management systems by companies in relation to their sites.
- systematically, objectively, and periodically evaluating the performance of such elements.
- providing information on their environmental performance to the public.

Although voluntary like 14001, it may prove vital for companies to conduct business within the EU, although European companies are unlikely to stipulate 14001 as a trade requirement unless there is a potential market advantage.

This last objective requiring public disclosure of a company's environmental performance has raised major complaints from outside the European Union. The United Kingdom's Environmental Protection Act of 1990 contains some, but less stringent disclosure requirements than EMAS. Additionally, EMAS requires a policy to cover methods for:

- assessing, controlling, and reducing a company's environmental impacts.
- managing energy, raw materials, and waste.
- production process changes.
- product planning (design, packaging, transportation, use, and disposal).
- determining environmental performance and practices of contractors and suppliers.[3]

This last policy requirement is already raising questions from industries in non-European Union countries regarding the potential requirement for a supplier or contractor to have some certified environmental management system in place before being allowed to conduct business in Europe or with a European-based company with sites in other parts of the world. This situation is already being faced by the automotive industry, and an EMAS-mandated environmental management system will most likely fall next on the power, manufacturing, and waste disposal industries.

The EMAS regulation is being recognized by some non-EU businesses as a very comprehensive specification for an environmental management standard and, thus, are choosing to use it as the reference point for developing their environmental management system instead of ISO 14001. It is expected that ISO 14001 will not affect EMAS, but the difficult choice for many companies will be whether to focus on ISO 14001 or EMAS. Many view ISO 14001 certification as a stepping stone to EMAS. The decision will come from a company's primary goal: *performance* (EMAS) or *conformance* (ISO 14001). Since ISO 14001 is much broader in scope and less prescriptive, it means that businesses already registered to EMAS have the major infrastructure in place for registration to ISO 14001 if they so choose to do so — some most likely will not do so. The German chemical industry, due to its tremendous potential impact on the environment, has been the primary driver of the Eco-Management Audit Scheme (EMAS) and has basically "shunned" ISO 14001 and any of the "bridging" documents developed to link EMAS and ISO 14001.

2.3.2   British Standard 7750

In September 1996, the 15 representatives to the European Union voted to accept ISO 14001 as the sole European standard for environmental management which became known as EN 14001, *Environmental Management Systems — Specification with Guidance for Use.* This decision has and will result in the phase out of all other national environmental management standards throughout Europe, including British Standard 7750. The impact BS 7750 had on the development of ISO 14001 cannot be ignored. Many firms worldwide who have chosen to be certified to BS 7750 as their environmental management system have chosen not to invest further on other standards such as ISO 14001. As a much more rigorous standard, however, registration to ISO 14001 may incur very little extra cost.

British Standard 7750 was issued in 1994 as a United Kingdom environmental management standard and is considered the "father" of ISO 14001. BS 7750 requires a business have the following:

- Policy
- Management reviews
- Management audits
- Management records
- Organization and personnel
- Environmental effects evaluation and register
- Objectives and targets
- Management program
- Management manual and documentation
- Operational control

2.4   NORTH AND SOUTH AMERICA

2.4.1   MEXICO

With the introduction of the North American Free Trade Agreement (NAFTA), a great deal of attention on the environmental problems in Mexico, as well as in Latin America in general, has raised concerns. Many industries in the United States are worried about "pollution export and import" across the southern United States border that will affect U.S. border communities. Thus, there has been pressure on Mexican industries to implement ISO 14001 from firms in the United States who worry that their images may be damaged if they continue to conduct business with "environmentally irresponsible" firms in Mexico, especially those along the border. Mexico, however, has made great strides in the area of environmental protection, and has had an environmental auditing standard which predates ISO 14001 — Mexico's efforts have not necessarily been in response to NAFTA or pressure from United States industries. Although NAFTA does include environmental provisions, Mexico has had strong public commitment and government concern for improving the environment that preceded NAFTA.

Mexico's voluntary environmental auditing program was introduced by the Federal Environmental Attorney General (Procuraduria Federal de Proteccion al Ambiente or PROFEPA) in response to a series of industrial disasters in Mexico which showed there was a serious lack of trained environmental and safety inspectors. The audit is a comprehensive EH&S review and includes worker protection (safety and occupational hygiene) and environmental releases (such as soil and groundwater contamination from prior practices). Additionally, there are verifications of Mexico's environmental, health, and safety regulations and adherence to international EH&S-related standards (i.e., OSHA, EPA, API, etc.).[4] As an example, in order to become and remain competitive in the worldwide market, Mexico's chemical industry adopted the principles of the Responsible Care Program®.

Due to the introduction of TQM in the late 1980s and early 1990s, Mexico's industry has become much more receptive to TQEM-based management approaches. In addition, Mexico's environmental legislation (primarily the 1989 General Law on Ecological Equilibrium) has focused on a prevention-based approach to environmental management. Of major concern, however, are the SMEs (small to medium-sized enterprises) which constitute more than 90% of businesses in Mexico, who are not as familiar with the concept of TQM much less familiar with ISO 14001.

Mexico played a very late role in the development of ISO 14001 and now feels that it has to play catch up. The reason for this is twofold: (1) the lack of support for ISO 14001 early on by Mexico's government and industry has resulted in Mexico being unable to provide proper input into the standards as they relate to potential trade issues; and (2) there has been a general resentment from most of the Latin American and to a minor extent, African nations who feel they were "excluded" from participating in ISO 14001 by the nations who dominated their development — the United States, the European Union, and Canada. Recently, however, the Mexican government and larger businesses have stepped to the forefront as they have begun to see environmental performance as strategic in their efforts to expand markets. In particular, the electronics industry in Mexico (due to the large export of manufacturing from the U.S. electronics industry) and other industries owned by U.S. firms along the U.S./Mexico border have begun to focus more energy on ISO 14000.

The Mexican government, through a unit of Mexico's environmental secretariat, the National Institute of Ecology (INE), has begun incorporating ISO 14000 into Mexico's environmental laws. Additionally, the Ministry of Environment, Natural Resources, and Fisheries (SEMARNAP) has restructured to incorporate all agencies that are environmentally-related and begun issuing standards for hazardous waste, air emissions, and wastewater discharge. Mexico has begun to position itself as a world-class environment leader.

2.4.2 BRAZIL

Brazil has come under consistent pressure to reduce the destruction of the Amazon rainforests and to implement the concept of sustainable development throughout its industries. Although sustainable development has become a major factor in Brazil, most of the major Brazilian industries are implementing ISO 14001 as a means of avoiding future nontariff barriers and to assist in promoting market share both nationally and internationally.

A third reason for implementation is to comply with Brazil's national environmental laws and regulations that are structured very similarly to those of the environmental regulations in the United States. The enforcement of them has been very relaxed and it is hoped that implementation of an environmental management system like ISO 14001, which requires regulatory compliance evaluations, will help companies to become more regulatory compliant. Of primary difficulty, however, for many Brazilian companies is lack of experience: it has been difficult for small to medium-sized enterprises to get into regulatory compliance much less attempt to implement ISO 14001 because they lack finances and experience and do not have the experience of other companies to draw upon. Much as was done to promote and implement ISO 9000, the Brazilian government will be offering financial support, such as tax incentives.

2.4.3 THE REST OF LATIN AMERICA

The rest of the countries in Latin America have very vague environmental regulations without any specific guidelines. Most of them have rudimentary environmental programs and only began to evaluate ISO 14001 in its later stages. As in Mexico and Brazil, major trade issues and growing public concern over environmental exploitation will continue to drive them towards programs such as ISO 14001 and the development of better defined environmental regulations. Chile and Colombia are two such nations who are now taking a hard look at ISO 14001 as a national standard much as they did with ISO 9000. Industries in Chile and Colombia have begun to draw upon their experiences with ISO 9000 to provide a foundation for potential integration of ISO 14001. Chile, in particular, adopted ISO 9000 as its national standard (NCH-9000) in order to increase and maintain market share. Its mining industry is crucial to that country's economic health and the Chilean government may require registration to ISO 14001. In Colombia, many firms are in the early stages of implementing ISO 14001.

Many are using their ISO 9001 certification procedure as a foundation to work from as they go through the environmental management system process.

Most of the remaining Latin American countries, such as Argentina, do not have a government system that forces a company to implement any kind of environmental program. Additionally, registration to ISO 9000 is almost seen as a last minute implementation in order to meet a customer deadline. However, due to an ever-expanding and increasing role in the global economy, Latin American countries are facing increased pressure to implement not only an environmental management system, but also to catch up with a quality management system (ISO 9000). With the exception of the larger international companies who already have the financial resources to implement ISO 9000 and ISO 14001, failure to do so could create negative financial and economic consequences for a nation.

2.4.4 CANADA

Canada has been one of the drivers of ISO 14001 primarily because of its significant potential impact on sustainable development (Chapter 3). Because of its large forest industry and fisheries, Canada has been at the forefront and a champion of sustainable development industries. Much of its "wealth" is based on its abundant natural resources and, as such, Canada has placed a high priority on the environment on federal, provincial, territorial, and local levels of government. A strong commitment to ecosystem management is now a priority in most corporate Canadian overall business decision-making.

Canada not only hosted the Montreal Protocol (Chapter 3), but is also the home of the International Institute for Sustainable Development (IISD) which plays a key role in monitoring and developing programs and policies that are concerned with integrating trade and the environment.

2.5   ASIAN/PACIFIC RIM

The rapid industrialization of the Asian/Pacific Rim has come at an extraordinary pace and has given Japan, Korea, Singapore, Hong Kong, and Taiwan a very modern and westernized style of living. This rapid growth, however, has not only brought prosperity, but it has also impacted their local environments. All of these nations have land constraints due to their size that in turn has created issues concerning overpopulation, available landfill space, and a depletion of natural resources that are already meager at best. These nations are dealing with air, water, and soil pollution, however, in a manner which is quite different from the United States and Europe — using corporate research and development rather than regulatory pressure. With the electronics industries leading the way, registration to ISO 14001 is growing at a rapid pace in this area.

2.5.1   JAPAN

Much like the rest of the world, the embracing of ISO 14001 has come from major manufacturing and electronic firms. Companies such as Toyota, Matsushita

Electric Industrial, Sony, Honda, and Asahi Chemicals have had strong environmental management programs in place for some time dating back primarily to the early 1970s. Most of the environmental management programs being implemented by these and other Japanese companies do conform to ISO 14001, despite the fact that ISO 14001 did not exist when implementation of environmental management systems began in the early 1980s. Because of the strength of their environmental management programs, several have indicated they will be seeking self-declaration rather than third party certification for ISO 14001.

Because of this, only a handful of large firms are embracing ISO 14001 and general knowledge of ISO 14001 throughout Japanese industry is very limited. In order to increase awareness and implementation of ISO 14001, the Japanese Ministry of International Trade and Industry (MITI) has stepped up a campaign requesting companies to prepare an ISO 14001-based environmental management system. This request is based on a realization that environmental problems extend across borders and can no longer be viewed as just a national concern. MITI is a government department which formulates industrial policy and its standards department promulgates national standards established by the Japanese Industrial Standards (JIS) Committee and disseminates international standards.[5]

This awareness and concern also has led the Japan Federation of Economic Organizations, also known as the Keidanren, to develop its Global Environment Charter (Appendix G). The charter has the basic philosophy that "a company's existence is closely bound up with the global environment as well as with the community it is based in." By using quality control techniques such as root cause analysis and continuous improvement, Japan has placed more emphasis on research and development techniques to deal with its environmental issues rather than placing heavy emphasis on issuing stringent regulations. Ensuring that environmental concerns are part of a product or process design review ensures that product stewardship becomes ingrained in their manufacturing and national philosophy.

2.5.2 AUSTRALIA

Australia became one of the very first countries to adopt ISO 14001 after it was elevated to Draft International Standard. By mid-August 1996, 30 Australian companies had become certified to either ISO 14001 or BS 7750. This same month, however, the Australian government decreed that companies seeking government contracts would no longer need to be certified to ISO 9000. This contradicts the plans of other governments who are stepping up requirements for ISO 9000 certification for the procurement process. This news is naturally expected to be extended to ISO 14001.

The Australian government has indicated that its intent was to give small to medium-sized (SMEs) companies a better chance at being awarded government contracts against the larger companies. As of early 1997, however, very few companies in Australia are implementing ISO 14001. One of the reasons behind this has been Australia's general view of ISO Standards — they are "rules" and not voluntary standards.[6]

2.5.3   HONG KONG

In contrast to Australia, the Hong Kong government has mandated that all consulting firms and contractors who conduct business with the government totaling >$1.25M must have ISO 9001/9002 certification with the expectation of the same requirement for ISO 14001 not far behind.

As in most of the developed areas of the world, Hong Kong's larger firms, based on their experience with ISO 9000, recognize ISO 14001 will be a requirement for conducting international business. Several of these large firms have been sending out questionnaires asking their suppliers for information regarding their environmental management programs, if any. It is expected that this information will be used to give suppliers who have environmental management systems in place preferential treatment for contracts and general business.

2.5.4   TAIWAN

As was discussed at the beginning of this section, Taiwan is faced with an ever-growing dilemma in regards to land restrictions. As the second most densely populated country in the world, Taiwan is constantly challenged with the question as to where to dispose of its waste. Economic prosperity has brought an increase in industrial waste, household waste, and an increase in automobiles and its resultant air pollution.

Taiwan created its own Environmental Protection Agency (EPA) in 1987 to deal with these problems and although enforcement has been strict, compliance from all sectors has been a tough battle. Apart from a few major industries, the incentive to prevent pollution on the island of Taiwan has not taken hold, especially among the general population. Due to the lack of general municipal landfills and incinerators, solid waste continues to pile up on street curbs and trash litters many streets, especially outside Taipei. It has been estimated that >75% of Taiwan's solid waste ends up in landfills — this could be reduced through recycling and management of household waste. But dealing with and managing the problem is a monumental task. The need to build more incinerators has been met with a "Not In My Back Yard" response and educating the general public on the need to sort trash and thus minimize landfill is not being accepted. Although ISO 14001 is being pushed through industry, it will not address the core of Taiwan's problem.

2.6   TOTAL QUALITY ENVIRONMENTAL MANAGEMENT

As was stated earlier in Chapter 1, Total Quality Environmental Management (TQEM) has come about as a result of applying the philosophy of Total Quality Management to Environmental Management. Total Quality Management has helped advance the integration of environmental issues into everyday business thinking and has shown business leadership that environmental management provides an opportunity and not a problem.

TABLE 2.4
TQM and EM Integration

Principle	TQM Focus +	EM Focus =	TQEM Focus
1. Management Responsibility	Sets an "example" and takes the lead	Management responsible by law for impacts	Concerned environmental citizen
2. Training	Skill improvement of worker enhances contribution to the business	Skilled worker has great impact on protecting the environment	Greater impact on business and less impact on environment
3. Process Yields	Waste is a "defect"	Reduce waste to protect the environment	Environmental impacts are "defects"
4. Continuous Improvement	Proactive approach to corrective action	Reactive approach to corrective action	Proactive evaluation of environmental impact
5. Structure and Responsibility	Cross-functionality	Focused responsibility	Cross-functional and wider focus of responsibility
6. Customer Satisfaction	Customer is always "right"	Compliance driven	Protecting community is the "right thing to do"
7. Employee Commitment	All individuals understand the goal	Some understand goal-"forced" commitment	All are committed due to personal concerns
8. Statistical Analysis	Root cause analysis and prioritization	Emissions and discharge monitoring	Analysis used to prioritize and minimize emissions and discharges
9. Customers	*Internal* — Employees *External* — Consumers; Distributors; Stockholders	*External* — Regulatory Agencies; Neighbors and Community	Customer is "everyone" and the surrounding environment

Table 2.4 shows how the principles of Total Quality Management plus Environmental Management equals Total Quality Environmental Management (TQM + EM = TQEM).

2.7 CONCLUSION

In contrast to the quality movement, the environmental movement has been received and embraced much differently. Even within the so-called "developed" countries, the attention given to the environment has been very diverse. From an environmental standpoint, "developed" nations can be separated into three subcategories: a fully *developed* environmental program (with its strict regulatory requirements such as in the United States and the European Union), a *developing* program (such as in Brazil where a regulatory framework is in place, but application and enforcement are difficult), and an *underdeveloped* program (such as in Taiwan where environmental attention is just now gaining ground).

Understanding your national and local environmental requirements can be quite a challenge for a quality manager and, depending on the management level, the international diversity can be overwhelming. It is hoped that the diversity of international programs will be pared down with the global implementation of the ISO 14000 series of standards. This will allow an international company to consider quality and environmental management systems integration as not just a good idea for an isolated operational unit, but good business sense throughout their organization — locally, nationally, and internationally.

3   Background on ISO 14000

3.1   INTRODUCTION

The previous chapter has demonstrated that environmental management systems have been in place for close to 90 years and began primarily with the petroleum and its allied industries. The United States and the Western European nations have been the primary drivers of ISO 14000, but concern for the environment on an international basis began several years before the International Standardization Organization began to take the initiative when it developed the international standard. Although Chapter 2 discussed the environmental movement, its primary focus was on industry and national programs. The focus of this chapter will be to provide the reader with a background on more of an international level and how various programs and organizations have influenced the development of the ISO 14000 Environmental Management Standards.

But first, it is important to understand an overlying principle which is becoming the heart of both national and international environmental management "thinking" — the concept of Sustainable Development. A basic knowledge of this principle will assist the reader in understanding the impact sustainable development has had not only on those programs I already have discussed, but will provide the groundwork for this chapter and the reasoning behind the overall quality and environmental integration as well. An understanding and a commitment to Sustainable Development is extremely critical to a business' long term future and must play a primary role in decision-making. ISO 14000 has already begun to demonstrate that it can provide the foundational concepts for a sustainable development policy and implementation plan at all levels of management and government.

3.2   SUSTAINABLE DEVELOPMENT

Sustainable Development is defined as "development that meets the needs of the present without compromising the ability of future generations to meet their needs" and came from the 1987 World Commission on Environment and Development (the Brundtland Commission). Sustainable development took another step forward internationally at the United Nations Conference on Environment and Development (UNCED) or Earth Summit 1992.

Its principles took shape under what is now known as "Agenda 21" (Table 3.1). This document deals with the actions governments, national and international organizations, and industry must take in order to achieve Sustainable Development. There are seven strategic concepts which came out of the Brundtland Commission which provide the foundation for establishing public policy based on sustainable development:

TABLE 3.1
UNCED Agenda 21

1. Social and Economic Dimensions	2. Conservation and Management of Resources for Development
Chapter	Chapter
2 International Cooperation to Accelerate Sustainable Development: deals with investment and debt issues	9 Protection of the Atmosphere: deals with climate change, stratospheric ozone depletion, and with trade, trans-boundary air pollution
3 Poverty: treated as both a cause and result of environmental degradation	10 Land Resources: issues affecting the planning and development of land resources
4 Changing Consumption Patterns: examines global imbalances in the patterns of production and consumption	11 Forests: addresses forest conservation and development programs
5 Demographic Dynamics and Sustainability: focuses on population growth	12 Desertification and Drought: action needed to halt and reverse rate of land degradation due to overuse or inappropriate use of range lands
6 Human Health: safe water supply, sanitation, adequate nutrition, and shelter are basic to human health	13 Sustainable Mountain Development: focuses on rapid deterioration of mountain ecosystems
7 Human Settlements: deterioration of the urban environment	14 Sustainable Agriculture: providing for an adequate food supply for the world population
8 Integrated Decision Making: consideration of environmental, social and economic factors are addressed jointly.	15 Biological Diversity: essential goods and services depend on maintenance of the variety and variability of genes, species, populations, and ecosystems
	16 Biotechnology: a vision of opportunities that exist to contribute to sustainable development
	17 Oceans: issues related to protection and sustainable use of oceans and seas
	18 Freshwater Resources: quality and supply while maintaining hydrological, biological, and chemical functions of ecosystems
	19 Toxic Chemicals: assessment of risks in the use of chemicals and illegal trafficking in toxic substances
	20 Hazardous Wastes: deals with waste minimization and recycling
	21 Solid Wastes: deals with minimizing wastes and maximizing recycling
	22 Radioactive Wastes: ensuring safe management, transportation, storage, and disposal

TABLE 3.1 (continued)
UNCED Agenda 21

3. Strengthening the Role of Major Groups	4. Means of Implementation
Chapter	Chapter
24 Women: ensuring full and equal participation in all development activities and in environment and development decision-making	34 Technology Transfer: conditions under which technologies should be transferred between countries for the benefit of the global environment
25 Youth and Children: developing a greater role	35 Science: role emphasizing institutions, research programs, etc.
26 Indigenous People and Their Communities: special people and their environment	36 Education, Public Awareness, and Training: role of formal and informal education
27 NonGovernment Organizations: providing a focus for community involvement	37 Capacity Building: encourages endogenous capacity building in developing countries
28 Local Authorities: each is encouraged to create its own Agenda 21	38 Institutional Agreements: focus on role of existing institutions and the changes in roles and objectives
29 Trade Unions: role in reorienting development policies and promoting cleaner production	39 Legal Instruments and Mechanisms: improving effectiveness of existing and future law on environment and development
30 Business and Industry: primary role of private sector in achieving balanced development	40 Information for Decision-Making: focus on "Bridging the Data Gap" and "Improving Information Availability"
31 Scientists and Technologists: improving communication between scientific and technological communities and the public	
32 Farmers: focuses on role as managers and custodians of natural resources	

- Reviving growth
- Changing the quality of growth
- Meeting essential needs for jobs, food, energy, water, and sanitation
- Ensuring a sustainable level of population
- Conserving and enhancing the resource base
- Reorienting technology and managing risk
- Merging environment and economics in decision-making

The overriding concept is that environmental protection must be considered as critical as social and economic issues. Much like the "fire triangle," where a fuel source, oxygen, and an ignition source are all required to start a fire, sustainable development will not happen if any one element of the triangle is missing (Figure 1). One cannot ignore the human (social and economic) side of the equation any more than it can ignore the environmental side. As mentioned in the chapter introduction, many are beginning to recognize a primary tool for sustainable development as being

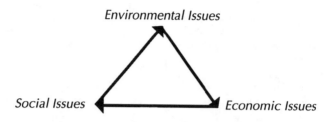

FIGURE 1.   The Sustainable Development Triangle.

the ISO 14000 Environmental Management Standards. ISO 14000 requires an evaluation of environmental aspects and those which would be considered "significant" aspects (or create an environmental impact) and, thus, causing a company to potentially investigate its supply chain due to the possible "influence."

By using the process of continuous improvement, a company would continually weave its economic considerations with its environmental aspects evaluations and, thus, drive towards a more efficient use of its natural resources. Additionally, sustainable development requires producers to focus on preventing an environmental impact before it occurs. A "cradle-to-grave" evaluation during a product or process review can minimize an impact on the environment and people. Many programs such as Product Stewardship, Design for Environment (DFE), Design for Manufacturability (DFM), etc. take some aspects of this into consideration. This focus overlaps into ISO 9001, **Element 4.4**, *Design Control*, and ISO 14001, **Element 4.3.1**, *Environmental Aspects*. More of this, however, will be discussed in later chapters.

3.2.1   INTERNATIONAL CHAMBER OF COMMERCE: BUSINESS CHARTER FOR SUSTAINABLE DEVELOPMENT

Known also as the "World's Business Organization®," the ICC was founded in 1919 as a nonprofit organization to serve the world's business community through the harmonization and promotion of trade, investment, and a free market system. The ICC has national committees throughout the world and has a major voice with the United Nations. With the introduction of ISO 14000 and due to its global influence, the ICC has begun working with the United Nations Environment Program (UNEP) to take an active role in promoting the environmental management standards.

In April 1990, the ICC issued its Business Charter for Sustainable Development. The charter, however, was due in large part to the work of a team of corporate environmental managers who make up the Global Environmental Management Initiative (GEMI). It was this group in fact which also developed and implemented the concept of Total Quality Environmental Management (TQEM) which was discussed in Chapter 2. The ICC Charter is supported by more than 2000 companies worldwide and has provided the foundation for industrial and national development of sustainable policies, environmental policies, and codes of business practice as it impacts the environment. The Charter contains 16 Principles for Environmental Management which are listed in Appendix D.

3.2.2 Earth Summit 1992

As the Sustainable Development movement gained momentum on an international scale, it became apparent that a more concerted effort needed to be developed to ensure its success. In June 1992, more than 170 countries met in Rio de Janeiro for the United Nations Conference on Environment and Development (UNCED) and what is now commonly known as "Earth Summit 1992." Representatives from these countries met to determine what the appropriate actions would be to ensure environmental preservation while at the same time ensuring economic development — the very concept of Sustainable Development.

With a focus on ever increasing global poverty and environmental destruction, the nations hoped to develop a much more organized effort in reversing this trend and to begin thinking as one world and synergistically through the elimination of environmental boundaries. Much like economics, the environmental impact of a nation extends well beyond its own national boundaries. The results of this meeting provided several nations with the impetus to develop or improve their national sustainable development programs. It not only raised the consciousness of the more developed nations, but helped the underdeveloped nations to recognize the level of environmental destruction they have been allowing within their own national boundaries. Industries coming in from the more advanced nations because of more "environmental freedom" and less stringent (or nonexistent) environmental regulations are now operating under closer environmental scrutiny and legal requirements.

3.3 MONTREAL PROTOCOL

Throughout the 1960s and early 1970s, some scientists had speculated that the earth's ozone layer was being destroyed by the use of chlorofluorocarbons (CFCs). It wasn't until 1974 that international attention began to focus on the issue, but the debate continued for over a decade with only a small percentage of the nations giving any attention to this issue. In 1985 an international agreement was finally signed which outlined global responsibilities for protecting the environment in addition to human health. This document, known as the Vienna Convention for the Protection of the Ozone Layer, however, did not impose accountability or obligations on any country — it was obvious that further work was needed.

Accountability and legal requirements were finally addressed in 1987 when the Montreal Protocol on Substances that Deplete the Ozone Layer was signed by 27 countries. It had subsequent amendments in 1990 and 1992 with more than 50 additional countries agreeing to abide with its requirements. This document addresses issues surrounding the stratospheric ozone layer by stipulating the phase out of certain ozone depleting compounds (ODCs), such as chlorofluorocarbons (CFCs), by Jan. 1, 1996, halons, carbon tetrachloride by 2000, and methyl chloroform (trichloroethane) by 2005. The agreement does not mean we will see CFCs go the way of the dinosaur in the near future. The parties to the agreement understood some CFCs have essential applications, such as medical, for which there are no known practical substitutes now available. Additionally, developed nations can continue to manufacture CFCs after 1996 during what is known as a "grace period" of

10 years (until 2006). With a sustainable development "mentality" taking hold on a global basis, however, the phaseout of ODCs has actually accelerated and major advances in chemical replacements have been remarkable.

3.4 CONCERNS OVER ISO 14000 STANDARDS

Ever since SAGE was formed by ISO to investigate the need for some international environmental standards, there has been increasing opposition to their passage on a variety of fronts and issues. Criticism has flowed freely from many underdeveloped nations and small-to-medium businesses as they have expressed primary concern over potential trade barriers, the cost of implementation, and their lack of applicability.

3.4.1 TRADE BARRIERS

One of the major questions being asked is whether or not ISO 14001 will become a requirement by customers of their supply chain, especially if conducting trade in the European Union and Japan, and it is uncertain in the near future whether the pressure to become third-party certified will increase. It is generally expected that most of the certifications to ISO 14001 will come from the European Union (primarily the UK), Japan and the United States. Because of this fact, the pressure from a customer to become third-party certified may not be as strong, especially in the major industries. In these cases, it will be market competitiveness which will drive certification. The only major concerns in these nations are if a government uses certification in purchasing decisions or bans non-ISO 14001 certified companies from selling within its jurisdiction which, however, would be a violation of the Technical Barriers to Trade (TBT) Agreement of the World Trade Organization (WTO). The real concerns lie with the underdeveloped nations and smaller industries in all nations.

In North America, immediate concerns stem from the North American Free Trade Agreement (NAFTA) which has turned the United States, Canada, and Mexico into a large free trade area with a principal objective of stimulating economic growth through equal access to each others' markets. A major requirement under NAFTA, however, is that Mexico must come up to speed on environmental issues as part of its participation (see Chapter 2).

3.4.1.1 General Agreements on Tariffs and Trade (GATT)

Founded in 1948, the General Agreements on Tariffs and Trade was initially an interim secretariat for the International Trade Organization (ITO). However, ITO faded away quickly and, for several decades, GATT continued to function as the primary framework for nations to ensure fairness among themselves in agricultural trade. One of the most significant developments to come out of GATT was the tenet known as Most Favored Nation (MNF) which allowed every member nation to be entitled to the same trade conditions that were applied to any other member's "most favored" trading partner.

GATT was, however, primarily an informal organization and suffered from several limitations. The major problem with GATT was that it did not have any "teeth" — it severely lacked any real authority to improve the global agricultural trade issues and was unable to deal with increasing trade issues among its member nations. Additionally, as an organization dominated primarily by Western nations, it included the majority of the world's economic wealth and strength, but did not represent the views of 75% of the world's population. After almost fifty years, the world's nations realized that a change was needed.

3.4.1.2   The World Trade Organization (WTO)

As the limitations of GATT began to have an effect on the global economy, new trade negotiations began to emerge in 1994. The culmination was the signing of the "Marrakech Protocol to the General Agreement on Tariffs and Trade 1994" which created the organization known as the World Trade Organization on Jan. 1, 1995. The WTO includes all of the GATT provisions plus some additional rules for trade in services and intellectual property and rules and procedures for settling disputes. The primary change from GATT to the WTO was in the binding authority given to the WTO over international trade issues with its member nations (totaling 120 members as of April 15, 1996).

What becomes significant about the WTO and the concerns raised with ISO 14001 is the link established between the WTO and Sustainable Development. The first paragraph of the "Agreement Establishing the World Trade Organization" states:

> Recognizing that their relations in the field of trade and economic endeavor should be conducted with a view to raising standards of living, ensuring full employment and large and steadily growing volume of real income and effective demand, and expanding the production and trade in goods and services, *while allowing for the optimal use of the world's resources in accordance with the objective of sustainable development, seeking both to protect and preserve the environment* and enhance the means for doing so in a manner consistent with their respective needs and concerns at different levels of economic development ...

In March 1996, the World Trade Organization established a Committee on Trade and Environment (CTE) with one of its primary goals to study the relationship between trade and the environment as they apply to promoting the 1992 Earth Summit's sustainable development policy. The CTE's foundational concern lies in resolving the conflicts arising from free trade and environmental protection. Many environmental organizations are already putting pressure on the WTO to include trade restrictions or sanctions on nations who are not complying with international environmental agreements such as the Montreal Protocol. The reverse situation has also been true — one of the WTO's first cases was its ruling that the stringent U.S. Clean Air Act discriminated against foreign oil refineries by forcing them to make cleaner gasoline. This, in the WTO's opinion, created an unfair advantage for the U.S. oil refineries. In August 1997, in fact, the U.S. EPA rewrote its rules on the cleanliness of the imported gasoline in order to settle the dispute. This is one example of the "influence" of the WTO and there is growing concern this type of pressure will be extended to ISO 14001 and EMAS.

3.4.2   DE FACTO REGULATIONS AND LEGAL ISSUES

This fear comes from many industries seeing trade sanctions being imposed through "*de facto* regulations." A "*de facto* regulation" is created when a government organization, such as the U.S. EPA, essentially incorporates references to an international standard in its regulations. An example of this would be if the U.S. EPA incorporated references to ISO 14001 in its governing of company operations — an argument could be made that, de facto, only ISO 14001 compliant companies are allowed to trade within the U.S. Evidence of this trend is already being felt in Europe with the continued implementation of the Eco-Management Audit Scheme (EMAS) in the European Union. For small to medium-sized companies, being banned from trade because of environmental sanctions would merely eliminate their source of income that could be used to implement the environmental system. It almost becomes a "Catch-22" situation.

Many firms have serious reservations about seeking certification due to other potential legal issues. Of primary concern is the confidentiality of EMS audits and whether or not they would fall under the protection of common law principles or other audit privilege laws. Another question being raised by the legal profession is whether or not the ISO 14000 series of standards have the potential to create what is known as a new "standard of care." What this means is that if the ISO 14000 environmental management standards become commonplace and a non-ISO 14001 certified company has an "environmental incident," they could potentially be found negligent in a court of law because they do not meet a customary standard of care. As with the trade issues, only time will tell.

3.4.3   COST OF IMPLEMENTATION

The cost to implement an ISO 14001-like environmental management system can be extremely expensive, especially if a nation or a small to medium-sized company is basically starting from "scratch." As a result of smaller financial resources many small to medium-sized companies have not given consideration to an environmental program — in terms of ISO 14001, this means no environmental policy, objectives and targets, understanding of environmental regulatory requirements, no continuous improvement efforts, and, above all, no senior management commitment. Most of their energy and financial resources are used to maintain any market shares they currently are managing to maintain.

Since small to medium-sized companies account for approximately 60 to 80% of the world's business and they generally have a very short life span, this could severely impede the desired change many want on global environmental protection. This concern is not unfounded — the cost to implement an environmental management system can be very cost prohibitive. The costs, however, can be reduced tremendously if an ISO 9000 or other quality management system has already been implemented — integration of an EMS can be possible.

3.4.4 APPLICABILITY

Developing countries have expressed anger over the ISO 14000 Standards because they were primarily developed by the larger U.S. and European industries. Because of their perceived exclusion, it is felt that the Standards do not adequately address the environmental concerns encountered in their countries. Most of this anger has come primarily from the Latin American nations who are now feeling tremendous pressure to get "up to speed." The pressures on Mexico through NAFTA have already been discussed in Chapter 2. With market economies rapidly changing, corporations in Latin America see the potential trade barriers looming on the horizon.

On the flip side, many Latin American governments see this as an opportunity for Latin American industries to finally comply with the long neglected environmental regulations. Unlike the United States and the European Union, who have more command and control environmental regulations, Latin American governments have had little command and control — the ISO 14000 Standards may provide the "bridging" system they need.

Much like the developing nations as a whole, small to medium-sized enterprises (SMEs) all over the world feel they have been deliberately omitted from any participation in the development of the ISO 14001 standards — as they are written they provide little assistance to SMEs in developing an environmental management system. Strong concern over the creation of potential trade barriers which will effectively lock SMEs out of markets and prevent them from growing economically, has forced ISO to begin addressing this issue by evaluating how the ISO 14000 Standards can be adapted for SMEs. As discussed in Section 3.4.1.2, the enforcement capabilities of the World Trade Organization has placed additional pressure to "even the playing field." As a result, an ISO 14001 bridging document for SMEs is being developed.

3.4.5 PRIOR STANDARDS

In Chapter 2 and in some of the Appendices, several lists of environmental principles and charters are presented. Many of the more advanced industries, particularly in the chemical industry, will not be pursuing certification to ISO 14001 because of the strength of their current environmental programs. The principles of the American Petroleum Institute and the Chemical Manufacturers Association, the standards of the Eco-Management Audit Scheme and British Standard 7750, and the various regulatory agencies are considered to be either comparative or superior to the ISO 14001 Environmental Management Standards. In some cases, seeking ISO 14001 certification may actually be a step backwards or provide no foreseeable improvement in the management system or in regulatory compliance. With this in mind, the cost of implementation becomes an additional deciding factor. For a large global chemical firm, the cost to seek ISO 14001 certification can run in the millions of dollars with very little net benefit to the overall program.

3.5 SIMILARITIES AND DIFFERENCES OF THE ISO STANDARDS

Appendix B, *Comparison of ISO 9001 and ISO 14001*, and Appendix C, *Related Sections of ISO 14001, ISO 9001, BS 7750, and EMAS*, show side-by-side comparisons of the various environmental standards. It is the correlation between the various elements and sub-elements of these standards which will provide much of the content for the remainder of this book with most of the comparison and subsequent integration occurring between the two ISO standards. The rest of this chapter will give a brief overview of their similarities and differences as a means of laying a foundation for the work ahead.

Table 3.2 lists the various standards of the ISO 14000 framework. Many of the standards at this point have not been officially released and, as mentioned, will not be considered in this book. Additionally, since the only document to be audited against will be the specification standard, ISO 14001, it is my intent to remain as focused as possible on providing information and a guideline that will assist you in certifying to ISO 14001.

Additionally, as we go through this process, we will be evaluating the various industry standards provided throughout this book that can be utilized as good resources for developing an environmental policy, objectives and targets, etc.

3.5.1 SIMILARITIES

It is to be expected that since ISO used the 9000 series as a foundation for developing 14000, there should be many areas which correlate to various degrees. It is this list of similarities which provides the intent of this book. Major correlations exist in the following elements:

- Policy
- Training
- Document control
- Management review
- Corrective action
- Records
- Internal audits
- Operation or process control

As you can see, this means that ~50% of the ISO 14001 elements can be integrated into an existing ISO 9001 structure. For some businesses who do not have ISO 9001 certification as of yet, but are first pursuing ISO 14001 instead, this means ~40% of ISO 9001 can be integrated into an existing ISO 14001 structure. This all depends on your business goals and primary focus. An in-depth evaluation of the two standards' elements can potentially result in an integration percentage that is much higher than initially perceived — approximately 60–70% can possibly be attained. It is realistic to assume that initially, the percentage will be much lower to start with and full integration will come after the groundwork has been laid.

TABLE 3.2
The ISO 14000 Series of Standards

Doc #	Description/Title	Purpose
14001	Environmental Management System (EMS)	Specification with Guidance for Use
14004	Environmental Management System (EMS)	General Guidelines on Principles, Systems, and Supporting Techniques
14010	Guidelines for Environmental Auditing	General Principles on Environmental Auditing
14011/1	Guidelines for Environmental Auditing	Audit Procedures
14012	Guidelines for Environmental Auditing	Qualification Criteria for Environmental Auditors
14013	Management of Environmental Audit Programs	
14014	Initial Reviews	
14015	Environmental Site Assessments	
14020	Goals and Principles of All Environmental Labeling	
14021	Environmental Labels and Declarations	Self Declaration Environmental Claims — Terms and Definitions
14022	Environmental Labels and Declarations	Self Declaration Environmental Claims — Symbols
14023	Environmental Labels and Declarations	Self Declaration Environmental Claims — Testing and Verification
14024	Environmental Labels and Declarations	Environmental Labeling Type 1 — Guiding Principles and Procedures
14025	Type III Labeling	
14031	Evaluation of Environmental Performance	
14040	Life Cycle Assessment	Principles and Framework
14041	Life Cycle Assessment	Life Cycle Inventory Analysis
14042	Life Cycle Assessment	Impact Assessment
14043	Life Cycle Assessment	Interpretation
14050	Terms and Definitions	Guide on the Principles for ISO/TC 207/SC6 Terminology Work

3.5.2 DIFFERENCES

There are several major differences between the two standards and, in some cases, integration is at best a longshot. The Environmental Management Standards contain some elements which are particularly unique:

- Environmental aspects
- Legal and other requirements
- Emergency preparedness and response
- Objectives and targets

Although these four elements are unique, you will see that all of them can be integrated into various sections of ISO 9001. What makes these elements unique is the fact that they drive a much more defined "continuous improvement" process as compared to ISO 9001. The need to "prevent pollution" (which is a specific policy statement requirement demanded by the U.S. EPA) is the foundational goal for implementing ISO 14001 — this means that a structured continuous improvement process has to be "ingrained" in the environmental management system to achieve this goal. You will see in Chapters 6 through 19 that integration on a large scale can be accomplished with a "little ingenuity" and plain common sense.

3.6 CONCLUSION

Only the passage of time can determine what the impact of ISO 14001, EMAS, or any other environmental management system will be on the world's industries. As the players in the game square off with each other, the lack of a "referee" to officiate will only result in long-term quarreling and a disregard of the rules. With their strong command and control environmental regulatory systems, the United States and European Union nations, in conjunction with environmental "watchdog" organizations, would like to see the rest of the world come up to the same level of skill in terms of the quality and concern for the environment. On the other hand, less developed nations, small to medium-sized enterprises, and organizations such as the World Trade Organization, have a primary focus of obtaining or maintaining a share of the economic market. The WTO's "split personality" to both support sustainable development and ensure a free trade system can only create more hard feelings on both sides. Requiring the more advanced nations to back down on some of their environmental restrictions in order to accommodate another nation or industry's need to participate in free trade could potentially undermine the overriding goal. The ultimate goal should be a win-win situation for all concerned, but I'm afraid that the current goal is for one side to win and the other lose.

Only time will tell…

4   Benefits of ISO System Integration

4.1   INTRODUCTION

In the past, a business' manufacture of a quality product was the consuming public's main concern, but this is now changing. The public in general is more concerned about the pollution which local industry is emitting to the air, drinking water, and soil — too many "Love Canals" have done nothing but fuel their concern. Additionally, with increasing requirements for both the public and business sectors to recycle their waste materials, environmental issues have now begun to command the most attention. Business stockholders and other stakeholders are also becoming more concerned about increasing litigation against companies through private individuals and regulatory agencies because of the severe impact on profitability and dividends.

As our economies continue to merge globally, fragmented systems can potentially create barriers for a company attempting to penetrate new markets or even maintain current markets in the long run. Without one fully integrated management system, businesses will face critical decisions concerning their survival and long term prosperity. When there is a diversity of management systems, it becomes very difficult to manage them — inefficiency and substandard product can result. Management must realize that a product manufactured in their own country may not be acceptable in another country because of quality and/or environmental "deficiencies." As you can see from the previous chapter, the weaving of sustainable development into the World Trade Organization's agenda will continue to cause business management to include and consider environmental issues as part of their overall business strategy.

Although Appendices B and C provide a very detailed comparison of various environmental standards, it is difficult to see the benefits that can be derived from the integration process. Figure 1 shows the goal of the integration process with the programs that potentially may influence a single system and the potential benefits coming out of it. The figure is not all inclusive, but merely gives you a snapshot. This chapter is intended to focus on more of the specifics.

4.2   COST EFFECTIVENESS AND PROFITABILITY

One of the first questions senior management will ask when confronted with a decision to implement ISO 14001 will be, "How much will it cost?" Like any other investment or expense, managers want to know what their Return on Investment (ROI) will be. A more appropriate question should be, "How much will it save?" Environmental professionals have tried to demonstrate to senior management that an EMS integrated into everyday business decisions can and will save money. Efforts

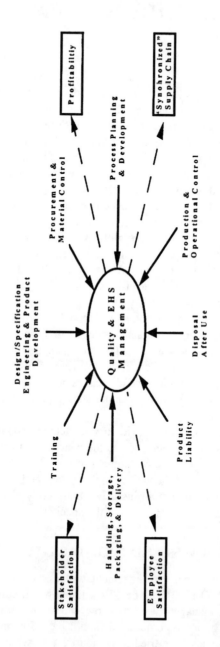

FIGURE 1. The goal of quality and environmental integration.

to prevent pollution, reduce solid waste generation, ensure legal compliance, etc. can have a significant impact on the efficiency of a process line by helping to identify process losses and, thus, improve yields.

An example of this might be where an environmental manager is attempting to eliminate an air emission's permit for a process. In working with a process engineer, it is seen that yields and quality fluctuate due to a poor soldering process. The product being manufactured uses an alcohol-based flux (e.g., a volatile organic compound) for the soldering operation that, additionally, must be run through a wash system with a cleaning solution. The environmental manager and/or engineer may determine that a water-based flux provides an excellent solder bond and that the washing system can now use warm water without a cleaning solution to provide a very clean, high quality product. The project has resulted in the elimination of an air emission's permit, the associated regulatory fees, and any logging or database requirements. Additionally, the purchase of the cleaning solution has been eliminated and the costs associated with its waste disposal. Another side benefit is the elimination of a potential regulatory noncompliance with the conditions required by the air permit. Thus, actual and potential savings have occurred. The basic premise behind this scenario is to demonstrate that operating costs can be reduced through sound environmental practices, primarily through the identification of an environmental aspect and determining how best to control or eliminate it.

In addition to the environmental aspects, there are other specific areas within the ISO 14001 standards that will contribute to cost savings. In particular, the development of objectives and targets can provide the driving force behind continual improvement and legal compliance. Utilization of an environmental review as part of a design review can minimize the impact of a new product or process.

Cost savings will also be evident if one realizes that integration into ISO 9001 will reduce the amount of man-hours spent in developing and implementing an environmental management system. With the ability to "piggy back" on an already existent ISO 9001 management system, the environmental manager can keep new program and procedural development to a minimum.

4.3   DOCUMENT CONTROL

As the ISO 14001 EMS was being drafted, one of the initial arguments which came out was the concern over the potential burden that would be placed on an already overloaded ISO 9001 document control program. This would, of course, be a valid concern if integration were not possible. Since the ISO 14000 standards were deliberately fashioned after the ISO 9000 standards, it is obvious that providing the ability to integrate was not far from the drafters' minds. The draft ISO 14000 guideline, in fact, states:

> Where elements of the EMS are integrated with an organization's overall management system, the environmental documentation should be integrated into existing documentation.

By developing and writing programs and procedures in a manner consistent with the ISO 9000 framework, an organization can very easily control its EMS document framework. Specifics of document control integration will be discussed in a later chapter.

4.4   INSURANCE

Several prominent members of the insurance industry have indicated that companies who pursue ISO 14001 or some other environmental management system may potentially see a reduction in their insurance costs. Insurance and industry representatives may see an EMS as a form of liability protection and may include special exclusionary language in insurance policies. Insurance companies pay out millions of dollars annually for coverage they provide for environmental pollution, legal penalties, lost revenues, and court and other litigation costs. An environmental management system shows the insurance carrier that a company is committed to its stakeholders: the insurance company itself, the regulatory agencies, the public, and company stockholders.

In order to put a price on an environmental management system, it has been recommended that insurance carriers put a "price tag" on environmental aspects and then prorate the level of "significant impact." By considering a company's list of significant impacts, an insurance carrier can potentially charge lower premiums. A downside to this, however, is the potential for providing legal compliance and any other information associated with environmental risks or audits.

4.5   AUDITING

The costs incurred to maintain compliance with ISO 9001 over the course of a three-year period potentially can run into the hundreds of thousands of dollars. By adding to this the potential costs for auditing an EMS, a chief financial officer will automatically become "gun shy." Because of the many expenses incurred for a third party certification for such things as document and report preparation, travel expenses, and the cost in man-hours for the auditee, etc., it's important to consider the benefits from conducting a joint QMS and EMS audit. For some large firms, the initial and triennial certification for just the QMS program can take anywhere from a week to two weeks. Due to its large size, the same firm can incur similar costs for an EMS audit. The combined auditing time can thus be anywhere from two to four weeks and may not include the time and expense incurred from mid-year surveillance audits! A joint QMS and EMS audit can save money in the areas already mentioned above, as well as making it much easier to prepare and train personnel to understand their responsibilities.

One of the critical issues concerning an integrated audit stems from potential complications that may arise from an audit team's confusion over the general architecture or structure of the joint systems. If integration of a QMS and EMS is in your plans, it is highly recommended that the integration take place over a period of time in order to minimize the initial disruption and to allow the joint system to have time to function smoothly. Only then should the joint audit be considered.

4.6 OVERALL BUSINESS DECISION-MAKING

With the preponderance of quality systems and ever-increasing environmental systems, it becomes even more important to integrate all of the systems into one synergistic business system. As nations and industries continue to adopt QMS and EMS systems and, potentially, require their supply chain to comply, it becomes critical that all of the various elements of a business be managed as a unified structure. Many businesses have become too fragmented with the "right hand not knowing what the left hand is doing."

Throughout this book I have attempted and will continue to show that good environmental management makes "good business sense." Over and over again, good environmental management has demonstrated its impact on improving operational and process control and, ultimately, on cost effectiveness and profitability. Personnel not only become aware of the impact their job has on quality, but also on the environment — they understand that the two cannot be separated. A waste generating process can become much more efficient and profitable if operators are aware that inefficiency can also increase pollution — the consequences of their mistakes can create product waste which, in most cases ends up in a landfill (i.e., it is also an environmental waste):

Product Waste = Environmental Waste = Reduced Profits

4.7 CONCLUSION

Implementing and integrating quality and environmental systems can have a significant impact on the financial success of a business. Financial institutions have begun considering the evidence of a sound environmental program as a potential credit evaluation tool. Lenders could be more willing to extend long-term credit and financing and give preferential treatment to a company if there is:

- a potential increase in market share access and acceptance.
- a reduction in costs.
- the attainability of a competitive advantage and faster time to market.
- a reduction in the costs on components and materials acquisition.
- a reduction in administrative and material expenses.

It will continue to become more evident that management must make environmental strategy as important as its marketing, financial, operational, and R&D strategies when establishing its short-term and long-term business objectives and targets — to include the environment as part of its vision and mission statements, as well as in its policy.

Part II

Integrating The Policy

5 The Policy

5.1 INTRODUCTION

Having spent time in the first four chapters laying down a foundation for the integration process, we can now begin to focus our attention on the actual process itself. In this chapter and the next several to follow, we will use the major ISO 14001 elements (Policy, Planning, Implementation and Operation, Checking and Corrective Action, and Management Review) as a base to be built upon as we go through this process. For instance, this chapter will use ISO 14001 **Element 4.2**, Environmental Policy, as the starting point and then see how various quality systems can be integrated with it.

Each of these chapters will be organized in a similar format which will include an evaluation of the various Quality Management System and Environmental Management System requirements, what a combined management system can or might look like, and what an auditor might look for if a joint QMS and EMS audit were conducted. Each requirement will be dissected in order to provide a better overall picture to view the integration process. Additionally, any differences that cannot be integrated will be discussed.

In Appendix B, I have provided a very detailed table that shows the cross-referencing between ISO 9001 and ISO 14001. This table will provide the primary foundation for the integration being discussed in the following chapters. In addition, Appendix C shows a comparison of the two ISO standards, British Standard 7750, and the Eco-Management Audit Scheme.

5.2 ISO 9001 REQUIREMENTS

The U.S. Automotive requirements for ISO 9001, QS-9000, will be evaluated primarily in conjunction with ISO 9001 and will, on occasion, be evaluated separately for specific QS-9000 requirements as they come up. In the case of establishing a policy, there are not separate requirements between the two standards. The requirements for a quality policy are found under Section 4.1, *Management Responsibility*:
Section 4.1.1 of ISO 9001 states:

The supplier's management with executive responsibility shall define and document its policy for quality, including objectives for quality and its commitment to quality. The quality policy shall be relevant to the supplier's organizational goals and the expectations and needs of its customers. The supplier shall ensure that this policy is understood, implemented and maintained at all levels of the organization.

In order to better understand what this requirement means and to ensure good compliance during an audit of an integrated policy, let's dissect it and highlight its various subcomponents:

> The supplier's management with executive responsibility shall define and document its policy for quality, including:
>
> * *objectives* for quality,
>
> and its commitment to quality. The quality policy shall be:
>
> * relevant
>
> to the supplier's organizational
>
> * *goals* and the
> * *expectations* and *needs* of its customers.
>
> The supplier shall ensure that its policy is:
>
> * understood,
> * implemented, and
> * maintained
>
> at all levels of the organization.

You will note that there are eight specific sub-requirements built into the overall **Element 4.1.1**. The reason for breaking this requirement down in such a manner will become evident when we look at ISO 14001's requirements.

5.3   ISO 14001 REQUIREMENTS

Element 4.2 of ISO 14001 states:

> Top management shall define the organization's environmental policy and ensure that it: (a) is appropriate to the nature, scale, and environmental impacts of its activities, products, or services; (b) includes a commitment to continual improvement and prevention of pollution; (c) includes a commitment to comply with relevant environmental legislation and regulations, and with other requirements to which the organization subscribes; (d) provides the framework for setting and reviewing environmental objectives and targets; (e) is documented, implemented and maintained and communicated to all employees; and, (f) is available to the public.

In a similar manner, let's dissect the subrequirements of this element:

> Top management shall define the organization's environmental policy and ensure that it is appropriate to the:
>
> * nature,
> * scale, and
> * environmental impacts,

and its activities, products or services; includes a commitment to:

- continual improvement and
- prevention of pollution;

includes a commitment to comply with:

- relevant environmental legislation and regulations, and
- with other requirements

to which the organization subscribes; provides the framework for:

- setting and reviewing environmental *objectives* and *targets*;

is:

- documented,
- implemented,
- maintained, and
- communicated

to all employees; is:

- available to the public

At first glance the requirements for ISO 14001 appear to be much greater than ISO 9001, when, in fact, they are very similar. Before we proceed with a side-by-side comparison of the two standards, it is important to discuss three very important differences found in ISO 14001.

5.4   POLICY DIFFERENCES

In the policy requirements for **Element 4.2** of ISO 14001, there are three items which do not appear in ISO 9001, but will have to be specifically included in an overall integrated policy: (a) the commitment to continual improvement; (b) the commitment to prevention of pollution; and (c) commitment to legal and other requirements. The inclusion of these in any policy, whether strictly environmental, quality, or operational in nature, must be fully understood, because of the ramifications to the resulting system and what auditors may end up evaluating as part of the audit scope.

5.4.1   CONTINUAL IMPROVEMENT

What is meant by continual improvement and what is its potential impact on your management system if included in a policy statement? Section 3.1 of ISO 14001 defines continual improvement as a:

process of enhancing the environmental management system to achieve improvements in overall environmental performance in line with the organization's environmental policy.

Additionally, Section 4.5.3 of ISO 14004, *Environmental Management Systems — General Guidelines*, states:

> The continual improvement process should ... identify areas of opportunity for improvement of the environmental management system which leads to improved environmental performance.

Although this may sound confusing at first, a closer inspection of these statements clearly shows that better environmental performance will be a result of the system's continual improvement process. Although this is an ISO 14001 requirement, there are practical benefits to be gained by a quality management system from a commitment to continual improvement. After all, what business doesn't want to continually improve its processes and the overall quality of its product? Although **Element 4.1.1** of ISO 9001 does not require this in the policy, you will see in a later section of this chapter, that it does contain foundational requirements to drive the continual improvement process. To use an analogy, both ISO 14001 and ISO 9001 policy statement requirements contain the "fuel" necessary to drive the continual improvement "engine." Thus, although the two standards differ in the requirement to state a "commitment to continual improvement," both also contain all of the necessary ingredients to improve performance through a continual improvement process. It is my opinion that the commitment to "continual improvement" in any policy can provide the basis for a better management system and will result in a significant impact on your operations' performance.

Although ISO 9001 does not contain this specific requirement, it is important to note that a continuous improvement "philosophy" is required under Section 2, *Sector-specific Requirements*, of QS-9000. Sector 2.1 states: "A comprehensive continuous improvement philosophy shall be fully deployed throughout the supplier's organization ..." Thus, continuous improvement becomes part of your basic organizational policy "by default."

5.4.2 PREVENTION OF POLLUTION

The inclusion of this requirement in ISO 14001 is one of the two primary differences from ISO 9001. However, as with the case of continuous improvement, QS-9000 is much more definitive in Section 4.9, *Process Control*. Again, prevention of pollution should become part of your organizational policy by default. ISO 9001 states:

> A supplier shall have a process to ensure compliance with all applicable government safety and environmental regulations, including those concerning handling, recycling, eliminating, or disposing of hazardous materials ...

The inclusion of this requirement was driven by the United States Environmental Protection Agency and created much debate over the difference between "pollution prevention" and "prevention of pollution" — this distinction must be clearly understood in order to ensure that the implementation of the management system clearly meets the ISO 14001 requirements. The U.S. EPA defines "pollution prevention" as "the use of materials, processes, or practices that reduce or eliminate the creation of pollutants or wastes at the source. It includes practices that reduce the use of hazardous materials, energy, water, or other resources and practices that protect natural resources through conservation or more efficient use." It's basic effort is to avoid the generation of pollutants throughout a process. In contrast, ISO 14001 defines "prevention of pollution" as "the use of processes, practices, materials, or products that avoid, reduce, or control pollution, which may include recycling, treatment process changes, control mechanisms, efficient use of resources, and material substitution." The standards are thus more concerned with how an organization manages its actual or potential pollutants. This particular definition won out because, as a generic international standard, many nations have neither the resources nor the capability in their manufacturing processes to avoid generating environmental pollutants — initial efforts have to focus on controlling the pollutants already generated and, hopefully, over time move towards "pollution prevention."

5.4.3   COMMITMENT TO LEGAL COMPLIANCE AND OTHER REQUIREMENTS

The third and final difference which must be addressed in the policy between ISO 14001 and ISO 9001 concerns compliance to regulations and any other requirements which your organization must adhere to (i.e., Responsible Care®, API, etc.). Although the policy requirements under ISO 9001 do not directly specify the need to commit to this, the standards as a whole do require a process to ensure legal compliance.

In ISO 9001, Section 4.9, *Process Control,* states:

> A supplier shall have a process to ensure compliance with all applicable government safety and environmental regulations, including those concerning handling, recycling, eliminating, or disposing of hazardous materials ...

Again, as in the case of continual improvement and prevention of pollution, legal compliance should become part of your organizational policy by default.

I would like to point out one major difference, however, between what ISO 14001, **Element 4.2**, *Environmental Policy*, requires and the above stated requirement under Section 4.9, *Process Control*, of ISO 9001. Specifically, Section 4.9 of ISO 9001 requires "a process to *ensure* compliance..." whereas ISO 14001's policy requirements are to "include a *commitment* to comply..." What ISO 14001 is *not* requiring is "compliance to regulations." In fact, as we look in a later chapter into ISO 14001, **Element 4.3.2**, *Legal and Other Requirements*, you will see that what is needed is a procedure to assist in identifying all legal and other requirements — it neither "guarantees" nor requires compliance. What we then face in writing the policy is a decision to include some statement which addresses the regulatory arena and any other standards, codes, etc. which you may or may not have to adhere to. Since the policy becomes the starting point for an auditor, it thus becomes critical to not include statements which are too "specific." It is highly recommended that the requirement under ISO 14001 be followed as written: "... a *commitment* to comply ..." One should avoid any definitive statements such as: "We will establish procedures to *ensure* compliance..." As one OSHA inspector told me: "You can never be in 100% compliance!" So don't commit to it in a policy!

5.5   OTHER STANDARDS AND PRINCIPLES

As you begin to think about what your organization policy will look like, it is highly recommended that you benchmark other organizations and standard-making bodies. Although you will have some business-specific requirements which you may want to include in your policy, there is no sense in "reinventing the wheel."

In some of the preceding chapters, I have provided some sources of information which can be used in your own policy. They include Deming's 14 Points for Management, API's Environmental Principles, and the Guiding Principles of Responsible Care®. In the Appendices you will find the International Chamber of Commerce's Charter for Sustainable Development, the Rio Declaration on Environmental and Development, and the Keidanren Global Environmental Charter.

Other bench-marking sources are other companies and, especially, those which have developed an "Operational" Policy as opposed to separate environmental and quality policies. An Operational Policy is essentially designed to show a business' overall management philosophy, core values, and mission in addition to satisfying some of the requirements of ISO 14001 and ISO 9001. But an Operational Policy can only be developed when a member from every facet of the business is involved.

5.6   DEVELOPING AN INTEGRATED POLICY

Let's now look at how we can actually assemble an "operational" policy which will satisfy the requirements of both ISO 14001 and ISO 9001. The best approach would be to first look at the two standards side-by-side and see where they match and then to write a statement that could be included in the policy. Table 5.1 shows a comparison of the two standards.

What needs to be done now is to begin formulating statements which: (1) reflect the common requirements; (2) address the requirements specific to ISO 14001; and (3) which can be edited in such a way as to generate a policy you would be satisfied with for public display.

The following statements can be used as examples:

- Our primary *goals* are to achieve complete customer satisfaction and to be recognized as an outstanding environmental citizen.
- Our primary *goal* is to meet the *expectations* and *needs* of our customers and the neighboring community through the use of effective *quality improvement teams* and *environmentally sound* management practices.
- We will achieve this goal by establishing *objectives* which will standardize, control, and *continuously improve* our operations.
- Our *goal* is to provide quality products, goods, and services which meet the *expectations* and *needs* of our customers through the use of *environmentally sound* management and technologies.

TABLE 5.1
Correlation of Policy Requirements

Element 4.1.1 — ISO 9001	Element 4.2 — ISO 14001
objectives for quality	environmental *objectives and targets*
relevant	*nature*
	scale
goals	environmental *objectives and targets*
expectations and *needs* of its customers	*other requirements*
understood	*communicated*
implemented	*implemented*
maintained	*documented*
	maintained
	continual improvement
	prevention of pollution
	relevant environmental legislation regulations and other requirements
	available to the public

- We shall provide products, goods, and services to our customers which are made (manufactured) to meet their *expectations*, and in a manner which meets all *regulations, industry standards, and other requirements.*
- Through the establishment of annual *objectives and targets* we will strive to *continuously improve* our operations.
- We will *continuously improve* so as to provide our customers with a quality product, to *prevent pollution*, and minimize the environmental impact of our operation.
- We will *communicate* this policy to all of our employees and make it *available to the public*, as a means to ensure their *understanding*, commitment, and active involvement in achieving our *objectives and targets.*

A combination of these statements or variations would meet the policy requirements found under the two ISO standards. With the exception of Akzo Nobel Chemicals, Inc., most companies do not have an integrated policy.

5.7   WHAT AUDITORS WILL LOOK FOR

The policy forms the foundation from which you build your management system and is, obviously, the starting point for an auditor. Besides the specific requirement to include *continual improvement, legal and other requirements*, and *prevention of pollution*, an auditor is mainly concerned about how the policy plays out through the system. Since a policy must be relevant to the nature and scale of your business, it is up to the auditor to determine if this, in fact, is so.

Some key things to remember: (1) make sure the policy is understood by your employees since their job functions must ensure compliance to it; (2) make sure the policy is available in as many ways as possible such as in training modules, rule booklets, on walls, at meetings, etc. — an auditor does not expect an employee to recite the policy by heart, but they must have a basic understanding of its contents and how it applies to their particular job function; and (3) make sure the policy is available to visitors, your neighbors, local agencies, etc. — use your customer service group as a vehicle to make it available.

5.8 CONCLUSION

In this chapter, my intent was not to write a policy for you, but to provide examples which can be incorporated into a policy that meets your expectations or is relevant to the nature and scale of your business.

If any part of your policy does not satisfy the ISO 14001 requirements, then the systems which are derived from it are lacking the vehicle which gives them "credibility." You can probably avoid "*hold*" points in other parts of the audit, but to encounter one right at the opening session could be very disheartening and take the "wind out of your sails."

In Appendix H, a figure which the interrelationship of the various ISO 14001 elements. You will note that the contents of your policy can be influenced by the ISO 14001 Standards themselves, but also by the applicable regulations, industry standards, codes of practice, your customers, the public, and so on. As you begin to formulate the contents of your operational policy, it is important to broaden the "scope of influences" on your business.

Part III

Planning

6 Environmental Aspects

6.1 INTRODUCTION

We now begin work on actually implementing the commitments that you have stated in your operational policy. In terms of the integration process, the area of Planning will perhaps be the most difficult to accomplish and may require a great deal of thought and ingenuity. What will make this difficult is the fact that the identification of a direct correlation between ISO 9001 (Section 4.2.3) and ISO 14001 (Section 4.3) is not as obvious in this area. Under ISO 14001, the Planning requirements take into account procedures to identify environmental aspects, legal and other requirements, and the establishment of objectives and targets. It is, in fact, the heart of the continual improvement process. This is not the case under the ISO 9001/QS-9000 framework.

In this chapter, we will look at the first major subsection of ISO 14001 — Section 4.3.1, *Environmental Aspects*. We will look at the comparative ISO 9001 sections and follow a similar process laid out in the previous chapter. Once you have gone through the certification process for ISO 14001, you will find as I did, that this section is perhaps the foundation of an entire audit. Coming to grips and having a well-developed understanding of Section 4.3 beforehand can remove a potential major stumbling block in an audit — the work you will do to satisfy this section will permeate throughout the rest of ISO 14001.

6.2 WHAT IS AN "ENVIRONMENTAL ASPECT"?

Before we can proceed further into this section of the chapter, it is important to understand what the definition of an "environmental aspect" is. In ISO 14001, Section 3.3, an environmental aspect is defined as an

> element of an organization's activities, products, or services which can interact with the environment.

The standard further requires the identification of a "significant" environmental aspect which is defined as an

> environmental aspect which has or can have a significant environmental impact.

What needs to be considered then are both *actual* and *potential* aspects and impacts. Each ISO 9001 section will be evaluated separately to see how the ISO 14001 "aspects" requirements can be integrated.

TABLE 6.1
Correlation of "Aspects" Requirements

ISO 9001 Section	Description
4.3.1	Contract Review — general
4.4.4	Design Input
4.4.5	Design Output
4.4.9	Design Changes
4.6.1	Purchasing — general
4.6.2	Evaluation of Subcontractors
4.6.4.2	Customer Verification of Subcontractor Product
4.7	Control of Customer-Supplied Product
4.13.1	Control of NonConforming Product — general
4.15.1	Handling, Storage, Packaging, Preservation, and Delivery — general
4.15.4	Packaging

6.3   ISO 9001 REQUIREMENTS

The correlation between ISO 14001, Section 4.3.1, *Environmental Aspects*, and ISO 9001 occurs in several areas. By reviewing the table in Appendix B, you can see that the aspects comparison to ISO 9001 occurs in the areas shown in Table 6.1:

6.3.1   DESIGN INPUT, OUTPUT, AND CHANGES

Before we begin evaluating how the standards are compatible, let's first see what ISO 9001 says and highlight the specific areas we are interested in:

Element 4.4.4 states: Design input requirements relating to the product, including applicable statutory and regulatory requirements, shall be identified, documented and their selection reviewed by the supplier for adequacy. Design input shall take into consideration the results of any contract review activities.

Element 4.4.5 states: … Design output shall … identify those characteristics of the design that are crucial to the safe and proper functioning of the product (e.g., operating, storage, handling, maintenance and disposal requirements).

Element 4.4.9 states: All design changes and modifications shall be identified, documented, reviewed, and approved by authorized personnel before their implementation.

Now let's review what ISO 14001 says:

Element 4.3.1 states: The organization shall establish and maintain (a) procedure(s) to identify the environmental aspects of its activities, products, or services that it can control and over which it can be expected to have an influence, in order to determine those which have or can have significant impacts on the environment.

The organization shall ensure that the aspects related to these significant impacts are considered in setting its environmental objectives.

What you may have been able to see is that a design review program can be an instrument for identifying environmental aspects. They may come about from the design of a new or modified process, product, or service and whether they can or potentially can cause a significant impact. An important element of a good design review program will be the inclusion of an environmental (and health and safety, as well) review of the new product, process, or activity. It is important that a design review team include an environmental manager who will have the knowledge and know-how to address the critical environmental issues.

The important environmental issues must, of course, address the requirements of ISO 14001. As the environmental review progresses, the design review team needs to address the following type of questions:

- Will this new or modified process or product create a new environmental aspect?
- Will this new environmental aspect create a significant impact?
- Will this new or modified process or product potentially change or influence a current environmental aspect?
- Will this new or modified process or product *potentially* change or influence a current significant impact?
- Will a contract with a current or future supplier create a new environmental aspect that may or may not create a significant impact?
- Will a contract with a current or future supplier *potentially* change or influence a current environmental aspect which may or may not create a significant impact?
- Are there any new regulations that must be taken into consideration with this new or modified product and/or process?

These are just a few examples that you may edit into your existing design review package that addresses an area known as Product Stewardship. The intent of a Product Stewardship program is to minimize or eliminate potential impacts of a product or material on the environment.

Additionally, it can address the health and safety hazards which may be experienced by manufacturing personnel or product installers (i.e., customers, etc.) — the ultimate goal is a "green product" in every way imaginable. You will find that throughout this particular chapter, the design review program will play a critical part in the integration process. Every area defined under the Planning requirements of ISO 14001 can be addressed through a good design review program.

Although ISO 9001 indirectly addresses environmental aspects, it is noteworthy to look at Section 19, *Product Safety*, in ISO 9004–1, *Quality Management and Quality System Elements, Part 1: Guidelines*. This guideline document very pointedly addresses Product Stewardship and the need to identify the various aspects of products and/or processes. This section states:

Consideration should be given to identifying safety aspects of products and processes with the aim of enhancing safety. Steps can include: (a) identifying relevant safety standards in order to make the formulation of product specifications more effective; (b) carrying out design evaluation tests and prototype (or model) testing for safety and documenting the test results; (c) analyzing instructions and warnings to the user, maintenance manuals, and labeling and promotional material in order to minimize misinterpretation, particularly regarding intended use and known hazards; (d) developing a means of traceability to facilitate product recall; and (e) considering development of an emergency plan in case recall of a product becomes necessary.

Before we leave this particular section, it is important to consider another area of EH&S aspects which could potentially be overlooked — the impact a facility may have. A design review program need not just address products or processes, but can also include new facility startups, structural modifications and upgrades, etc. with the result of a new process and/or product requiring the addition of a new exhaust system, an air conditioning system addition or upgrade, and so on. The consideration of environmental aspects now extends to the actual or potential creation of several other issues: (a) community noise from an exhaust system; (b) the requirement to utilize a refrigerant which may or may not be on a phase out list; and/or (c) an increase in the building's power, water, and gas consumption.

6.3.2 PURCHASING AND CUSTOMER-SUPPLIED PRODUCT

Element 4.6.2 of ISO 9001 states: The supplier shall: (b) define the type and extent of control exercised by the supplier over subcontractors. This shall be dependent upon the type of product, the impact of subcontracted product on the quality of final product and ...

Element 4.6.4.2 of ISO 9001 states: Verification by the customer shall not absolve the supplier of the responsibility to provide acceptable product, nor shall it preclude subsequent rejection by the customer.

The application of ISO 14001, **Element 4.3.1**, now takes on more of a quality "flavor." This may not seem very obvious at first, but let's consider the potential impact the *activity* of material purchasing can have on your operation and how it can *influence* or potentially influence the environment. Product specification, according to **Element 7.2** of ISO 9004, should include "performance characteristics (e.g., environmental and usage conditions and dependability); ... applicable standards and statutory requirements; ... packaging." It is obvious that poor product can: (a) result in scrap, low yields, etc.; (b) cause a tremendous loss of time in terms of processing the product from the time it is received into material control on through the final quality control process; and (c) a potential loss of other processing material and products which may be used in conjunction with the defective product. The processing of the product may also employ various chemicals which result in air emissions and hazardous waste — the more scrap, the more in-process chemicals you may use.

Another aspect to consider is the nature of the product itself. A good supplier and material evaluation program in conjunction with a design review can ask some of the following questions:

- Is it a chemical that is classified as hazardous (ignitable, reactive, toxic, corrosive, etc.)?
- Will the processing of this material create a hazardous substance which may be harmful to employees working with it?
- Will the processing of this material create a hazardous air emission?
- Is any part of this material potentially banned or restricted in another country where the final product is intended to be sold?
- Will "pass through" labeling requirements be in effect if an ODS is used?
- Will this use of a particular material in your product potentially subject your company to "take back" requirements with a customer?
- Is the material an ozone depleting substance (ODS) that is already banned or is it scheduled to be phased out in the near future?
- Will ultimate disposal of this material be subject to Land Disposal Restrictions (LDR)?
- Will there be special packaging requirements for this material resulting in the inability to recycle or reuse packaging?
- Will your supplier be using an ODS in the product manufacture which will require you to be in compliance with ODS "pass through" labeling requirements?
- Will the product require special packaging that may potentially come under special packaging and recycling legislation (particularly in Europe)?
- Will the nature of the product or material potentially result in special transportation requirements over land, air, or sea?
- Can a less toxic material/chemical be used?

As you can see, the evaluation of purchasing/material control must play a key part in the overall product development process. It is critical that materials be evaluated as early as possible during product or process design to ensure the type of questions above do not go unanswered. If this evaluation is not conducted, the potential impact can prove to be disastrous not only for yourselves, but for your customers as well.

The inclusion of both a material and a supplier evaluation during a design review for a new or modified product or process can go a long way in avoiding potential litigation, harming your company's reputation and goodwill, and/or suffering financially through process losses or lost/canceled orders.

6.3.3 CONTROL OF NONCONFORMING PRODUCT

If you consider your suppliers during the qualification process to be potential environmental "aspects" in addition to your ISO 9001 quality requirements, you can minimize or eliminate the product or material's actual or potential (significant) environmental impact.

Element 4.13.1 of ISO 9001 requires

... that product that does not conform to specified requirements is prevented from unintended use or installation" and "... shall provide for identification, documentation, evaluation, segregation (when practical), disposition of nonconforming product, and for notification to the functions concerned.

There is little else to say in light of the discussion just done above, but this section should be used to drive home the importance and the impact a quality program can have on the environment. The receipt and acceptance of a poorly manufactured product or material from a supplier can have a profound effect on your ensuing manufacturing operations that ultimately, of course, create various impacts on the environment. What you have is:

Poor quality → *Poor yields* → *Waste* → *Environmental Impact*

6.3.4 HANDLING, STORAGE, PACKAGING, PRESERVATION, AND DELIVERY

What we have again in this case is a program and aspect of an operation which management *"can control and over which it can be expected to have an influence..."* and which is focused on the impact of material flow from receipt to shipment.
Element 4.15 of ISO 9001 contains the key requirements to

> provide methods of handling product that prevent damage or deterioration; ... use designated storage areas or stock rooms to prevent damage and deterioration...; ... control packing, packaging, and marking processes...; and ... arrange for the protection of the quality of product, ... this protection shall be extended to include delivery to destination.

Material control personnel regularly handle materials which, by their very nature, are classified as hazardous. The handling, storage, packaging/preservation, and delivery of such materials becomes an environmental aspect of the overall operation. Let's consider a hypothetical example of a flammable solvent that is part of a final product shipped to a customer:

> The receiving department accepts shipments of 1/2-liter (500 cc), glass bottles of isopropyl alcohol. This is considered an environmental aspect with a potential impact because of its flammability, its volatility (air emissions), and its container (glass bottle). Because of its flammability, the bottles must be segregated into a special raw material storage room and, depending on the volume, placed into special storage cabinets (if total volume exceeds 120 gallons). Grounding straps must be provided to eliminate accidental ignition from static electricity. Additionally, the flammable storage cabinets must be vented to the outside atmosphere. Open containers or a broken glass bottle (or multiple broken bottles in the event of an earthquake — depending on your facility location) thus causes an air emission. The impact we are considering is the potential for a large structural fire if storage controls are not maintained. For a large volume of chemical in storage, this might be considered as a potential "significant" impact. What about the handling itself? If receiving personnel handle a large volume of this chemical, there could be an increase in the probability for a container or several bottles to be dropped. A chemical spill may result and the potential lack of spill control may allow some of the isopropyl alcohol to flow into a sewer or stormwater discharge source. Again, should this be considered a potential "significant impact"? What about delivery or shipping? Let's assume these 1/2-liter bottles of isopropyl alcohol are packaged into a kit with other components. The

inclusion of these bottles now requires the consideration of special packaging requirements under your appropriate transportation regulations. Let's also assume the transportation route of the truck is on a major, heavily traveled highway and it happens to become involved in an accident. Many of the bottles of isopropyl alcohol are destroyed with liquid now dripping off the truck and onto the roadway (fortunately, however, you have an emergency response plan in place for responding to such a scenario — did you consider this?)

I know this sounds somewhat unrealistic, but my intent with this scenario is to help you think beyond the "obvious" evaluation of your environmental aspects. All of the situations presented above may not be present or you have not experienced, but some pieces are most likely an everyday part of your operation. You may not at first have considered them an environmental aspect much less cause an actual or potential environmental impact. When you evaluate your products, raw materials, and other supplies as actual or potential environmental aspects, you may need to "dig deep" before calling them a "significant" impact.

6.4   WHAT AUDITORS WILL LOOK FOR

When you begin to consider what environmental aspects are present in your operation, you must not only be aware of their *actual* impacts, but also what their *potential* impacts may be. This is a key point for which an ISO 14001 will not overlook — the scenario above is not unrealistic in their eyes. One other point must be made and it is critical — think about the "whole forest and not just the trees!" It is important to not only identify and evaluate your environmental aspects within the context of your business operation, but to consider how they will impact your neighbors and the eco-system around you:

- Do you have equipment outside your building that may create "community noise?"
- Is your operation near a delicate eco-system such as wetlands, rivers, streams, bays, etc.?
- Do you use certain chemicals that have special reporting requirements such as prescribed under the Emergency Planning and Community Right-to-Know Act (EPCRA)?
- Does your air conditioning system use ozone depleting substances?
- Are you consuming various natural resources (power, water, gas)?
- Are there other natural resources, such as liquefied gases, which you are consuming?
- Do you transport chemicals through the surrounding neighborhood or utilize various transport systems such as air, rail, and sea?

One other important point — don't forget to consider the actual or potential influence of your "in-house" service organizations. This may include a site services group that manages your utilities, chemical waste program, etc. You may have allocated charges to your organization to pay for their services and, thus, they are no different from any other contractors you have on retainer or have a joint venture contract with. Additionally, Appendix H shows other sources for consideration in evaluating and identifying your environmental aspects: vendors, contractors, a design review, purchasing, regulations, standards, codes, customers, the public, and so on.

7 Legal and Other Requirements

7.1 INTRODUCTION

In today's world our legal and regulatory structure is making more and more of an impact on our daily lives. Although we are mainly impacted by laws which affect us on a personal level, the growing impact of environmental regulations is increasing. We can no longer drive our car without getting a smog certification, and used motor oil must be taken to a recycling center or picked up at the curb of our house. Environmental regulations, however, are being directed primarily at industry and, thus, can have a significant impact on your operations. It makes sense, therefore, to take heed and have a system in place to ensure your company is in compliance with environmental regulations and other requirements.

7.2 WHAT ARE "LEGAL AND OTHER REQUIREMENTS"?

In most cases, "legal" requirements are in reference to a company's commitment and obligation to comply with federal, state, and local environmental regulations. Specific examples may be to comply with air permits, water discharge permits, licenses, and other local ordinances.

Element 4.3.2 in ISO 14001 states: "The organization shall establish and maintain a procedure to identify and have access to legal and other requirements to which the organization subscribes directly applicable to the environmental aspects of its activities, products, or services."

The definition of "Other Requirements" is rather broad in scope and may include some of the following:

- Industry codes (i.e., Responsible Care® and STEP)
- Corporate policies and standards
- Partnership agreements
- National or international charters (sustainable development)
- International standards (i.e., ISO, BS, EMAS, etc.)
- Agreements with Public Authorities (i.e., EPA's Project XL)
- Private Codes (i.e., ASTM, ASME, NIST, and ANSI)

It is important to note any or all of the above (in addition to environmental regulations) should be considered as potential sources for identifying environmental aspects. Whether or not they are "significant" is dependent on the regulatory requirements, the (risk) assessment tool employed to define "significant," and any other methods available.

TABLE 7.1
Correlation of "Legal and Other" Requirements

ISO 9001 Section	Description
4.4.4	Design Input
4.4.5	Design Output
4.9(c)	Process Control
4.15.4	Packaging

7.3 ISO 9001 REQUIREMENTS

The correlation between ISO 14001, Section 4.3.2, *Legal and Other Requirements*, and ISO 9001 also occurs in a few sections. By reviewing the table in Appendix B, you can see that the legal comparison to ISO 9001 occurs in the areas shown in Table 7.1. You will note that these elements are included in Table 6.1 and have their foundation primarily in the design review program. It is important to note right up front, however, that if you rely entirely on the design review program to satisfy all of the requirements under ISO 14001, Element 4.3.2, you may potentially find yourself with a "*hold*" point at the end of an audit. ISO 14001 requires you to develop and have a *procedure* in place — a document which details the purpose, scope, responsibilities, references, and processes you plan to utilize in evaluating legal and other requirements. This is a critical point to remember! The standard specifically states:

> The organization shall establish and maintain a procedure to identify and have access to legal and other requirements to which the organization subscribes, that are applicable to the environmental aspects of its activities, products or services.

This requirement makes it extremely difficult to identify an ISO 9001 document which can have the ISO 14001 requirements integrated with and, in most cases, this will not be a possibility — a separate procedure will most likely need to be written to cover the ISO 14001 Standard.

A thorough environmental review during a design review can only provide a mechanism to ensure legal aspects are addressed for a new/modified product or process. It is, therefore, a mechanism to ensure compliance and avoid issues in the arena of product liability and product stewardship. For the purposes of this chapter, however, I will focus on the design review program as an excellent vehicle to continually focus attention on the legal requirements.

Let's now look at what ISO 9001 says:

Element 4.4.4 states: Design input requirements relating to the product, including applicable statutory and regulatory requirements, shall be identified, documented and their selection reviewed by the supplier for adequacy.

Element 4.4.5 states: ... Design output shall...identify those characteristics of the design that are crucial to the safe and proper functioning of the product (e.g., operating, storage, handling, maintenance and disposal requirements).

Element 4.9(c) states: The supplier shall identify and plan the production, installation and servicing processes which directly affect quality and shall ensure that these processes are carried out under controlled conditions. Controlled conditions shall include the following: ... (c) compliance with reference standards/codes, quality plans and/or documented procedures...

All of these elements are very clear about the need to include "statutory and regulatory requirements," "compliance with reference standards/codes," and "other requirements" as part of your review process and process control. The question then becomes how one can structure a design review to meet these requirements.

7.4 AN ENVIRONMENTAL DESIGN REVIEW

As you will see in the following subsections, the evaluation of environmental "legal and other requirements" during a design review must reach beyond your own national boundaries if you intend on manufacturing a product to be sold internationally or establishing a manufacturing process in another country. With the world continuing to "shrink" both economically and regulatory-wise with the continuing acceptance of international standards, a design review must broaden its scope of application.

7.4.1 NATIONAL REQUIREMENTS

For the purposes of this section, let's assume "national" refers to the legal requirements regulated by the United States Environmental Protection Agency (EPA). In this section, I will provide some questions that address specific requirements under the major environmental legislative acts. An abbreviated list of some of this legislation is found in Table 2.2. My purpose in this section is not to analyze the regulations in detail, but to provide some questions that may impact your review.

The Toxic Substances Control Act (TSCA) should be a primary legislation addressed due to the ramifications it has on chemicals entering from outside the country. When evaluating TSCA, the following types of questions should be asked:

- If a new chemical or substance is to be manufactured as part of this product design, has a Pre-Manufacturing Notification (PMN) been filed with the EPA?
- Is a chemical or substance to be used in the new or modified product going to be used in a significantly new way (Significant New Use Rule or SNUR)?
- Are all chemicals or substances currently on the TSCA Inventory List?
- Is any chemical or substance currently under evaluation by the EPA's Genetic Toxicology Program, by the NTP or IARC or listed on the Chemicals on Reporting Rules (CORR)?
- Is any chemical or substance to be used in a new or modified product intended to be imported into the country?

The Resource Conservation Recovery Act (RCRA) is concerned with the "cradle-to-grave" handling of chemicals, but focuses most of its enforcement on issues surrounding waste and its ultimate disposal. When evaluating RCRA, the following types of questions should be asked:

- What are the characteristics of the waste (e.g., toxic, reactive, ignitable, or corrosive)?
- What is the most effective means for disposal (i.e., landfill, incineration, etc.)?
- If the disposal option is for a landfill, will a leaching study need to be conducted to ensure none of the chemical(s) leaches into groundwater?
- Considering the volume of waste to be generated, will this classify the company as a "Small Quantity Generator" or a "Large Quantity Generator" (if not already determined)?
- Will a new hazardous waste profile result from the manufacture of this product?

The Clean Water Act (CWA) is concerned with "all waters" within the confines of the United States borders. It regulates the environmental pollution of our rivers, lakes, streams, bays, and offshore waters. When evaluating the CWA, the following types of questions should be asked:

- Will any process water contain heavy metals that may not exceed certain concentrations?
- Will any process water contain concentrations of organic materials that may alter the Biological Oxygen Demand (BOD) of the water and thus impact the sewage treatment system?
- Will any process water that is discharged into a sewer create the need for a new discharge permit or potentially alter an existing permit?
- Will there be any potential water discharge directly into the soil and which may impact groundwater?
- Are any chemicals classified as "Priority Pollutants"?

The Clean Air Act (CAA) regulates the emission of any chemicals into the air. When evaluating the CAA, the following types of questions should be asked:

- Will the process create the emission of a Volatile Organic Compound (VOC)?
- Will the process create the emission of a particulate?
- Will the process create the emission of a PreCursor Organic Compound (POC)?
- Does the process manufacturing the product already have a permit-to-operate or will it require the request for a new source permit?
- Will the emissions potentially alter the existing permit conditions?
- Are any of the chemicals used in the manufacture of the product classified as a Hazardous Air Pollutant (HAP)?

- Are any of the chemicals used in the manufacture of the product classified as a Criteria Pollutant?
- Will the manufacture of the product require the use of an Ozone Depleting Substance (ODS) which is included on a future phase out list?
- Will the use of a certain chemical require the need to install abatement equipment?
- In all of the cases listed above, can a less hazardous chemical be used?
- Will any of the chemicals create a change in any reporting requirements under Title V?

The Superfund Amendment and Reauthorization Act (SARA) is a part of the Emergency Planning and Community Right-to-Know Act (EPCRA). Both of them are concerned with the issues of controlling the release of a certain list of hazardous chemicals and providing the means to inform the public of any potential danger in the event of an accidental release. When evaluating the SARA, the following types of questions should be asked:

- Are any of the chemicals required to be reported on the Toxic Release Inventory (TRI)?
- Are any of the chemicals classified as Extremely Hazardous Substances (EHS)?
- Is there a Threshold Planning Quantity (TPQ) for any of the chemicals?
- In the event of releases outside company premises, is there a listed Reportable Quantity (RQ)?
- Will the new product and/or process require community notification (EPCRA)?

In the section just completed, I have attempted to give you a small sample of the types of questions that may raise a "red flag" during a design review. Additionally, most of these may result in the flagging of a potential environmental aspect that also happens to be a legal aspect as well. You may also notice that some of the questions on legal compliance may lead you directly to the identification of some significant impacts.

7.4.2 INTERNATIONAL REQUIREMENTS

The legal and other requirements that focus on the international level must be taken as seriously as your respective national requirements, especially if your firm sells on an international level. This part of the review can potentially be much more extensive than that incurred when evaluating your own regulations if your intent is to either manufacture and/or sell a product in another country. Additionally, with several nations now adopting ISO 14001 as their national environmental standards, your design review must take into consideration this potential aspect.

Let's look at a host of questions that can highlight areas of concern that may need to be addressed during a design review:

- Are any chemicals or materials listed in another country which are either totally banned or restricted?
- Are there any other special regulations directed at any of the chemicals or materials you intend on using?
- If the manufacture of the product is intended to be done in another country, what are the environmental regulations concerning air, water, soil, hazardous waste, solid waste, etc. in that country (they may be much more restrictive)?
- If a chemical or material is manufactured in another country, but will be shipped back into your country, what are the import requirements and the effect on any environmental regulations?
- If a chemical or material is to be imported or exported either by air or sea, what are the specific air and maritime transport requirements?
- If used in another country, what are the specific "end-of-life" issues which must be addressed (i.e., returned to manufacturer for final disposition, reused, recycled, etc.)?
- Are there any special packaging issues which must be addressed (especially in Europe)?
- Will the chemical, material, or product require special product hazard communication to the customers?
- Are there any special eco-labeling requirements (e.g., ODS/CFCs, hazard communication, product liability, etc.)
- Will there be any additional testing required to satisfy special consumer issues such as stipulated by the U.S. Consumer Product Safety Commission?
- Will a product require the inclusion of a Material Safety Data Sheet even though it may be classified as an "article"?

These questions have shown that we cannot be too narrow-minded in an environmental review during the design and development of a new product or process. A product intended to be manufactured or sold in several areas on an international level can create a very extensive and tedious environmental review.

7.5 WHAT AUDITORS WILL LOOK FOR

It is very important that an auditor not focus on a detailed compliance audit, but instead focus on whether there is a process or system in place that assists you in maintaining compliance — the standard requires that there be a procedure in place. The scope in this case is:

$$Policy \rightarrow Procedure \rightarrow Process$$

A good auditor will look for three primary things:

1. Is there objective evidence of a process being used to identify the requirements and how effective is the process?
2. Does the organization show an awareness of the legal requirements and the ability to maintain an awareness of the constant legal changes, and maintain access to them?
3. Were legal and other requirements considered in identifying aspects and potential significant impacts (as discussed in the previous sections)?

One of the foremost things to be aware of is that a qualified auditor must be familiar primarily with the relevant legal requirements and to a lesser degree, with the "other" requirements. It becomes important, therefore, for you to address this issue before an auditor shows up at your door. It does not make good sense to allow an auditor who is potentially uneducated in your relevant legal requirements, have no issues found in his revoew of this element, and then to have another auditor during a surveillance audit find a major problem.

8 Objectives, Targets and Environmental Management Programs

8.1 INTRODUCTION

This section will combine two elements of ISO 14001: 4.3.3, *Objectives and Targets*, and 4.3.4, *Environmental Management Programs*. The reason for this is the progression evidenced from one element to another and, for both of them, a natural progression from Element 4.3.1, *Environmental Aspects*:

Aspects → Objectives → Targets → Programs

8.2 ISO 14001 REQUIREMENTS

The requirements for establishing objectives and targets and programs (projects) are listed here, but I would like to point out the particular attention you must give to the things that must be "considered":

Element 4.3.3 states: The organization shall establish and maintain documented environmental objectives and targets, at each relevant function and level within the organization. When establishing and reviewing its objectives, an organization shall *consider* the legal and other requirements, its significant environmental aspects, its technological options and its financial, operational and business requirements, and the views of interested parties. The objectives and targets shall be consistent with the environmental policy, including the commitment to prevention of pollution.

Element 4.3.4 states: The organization shall establish and maintain a program(s) for achieving its objectives and targets. It shall include (a) designation of responsibility for achieving objectives and targets at each relevant function and level of the organization; (b) the means and time-frame by which they are to be achieved. If a project relates to new developments and new or modified activities, products or services, program(s) shall be amended where relevant to ensure that environmental management applies to such projects."

Before we continue I would like to take this time to briefly define what is meant by a "program" under the ISO 14001 Standards. The authors of the standards have defined it as that part of the environmental management system that addresses the scheduling, resources, and responsibilities for achieving the objectives and targets. It identifies the specific activities dealing with individual processes, projects, products, services, sites or facilities within a site (see ISO 14004, *General Guidelines on Principles, Systems, and Supporting Techniques*).

TABLE 8.1
**Correlation of "Objectives, Targets,
and Programs" Requirements**

ISO 9001 Section	Description
4.1.1	Quality Policy
4.1.3	Management Review
4.2.3(b)	Quality Planning
4.14.3(b)(c)	Preventive Action

The reason we are taking the time to define it is because of the potential confusion that may be encountered with an auditor. In many circles, the ISO 14001 definition of "program" is more closely compared to a "project" which is considered to have a definitive start and end — when the target is achieved, the project is closed out. If you have this context in mind, then "program" is more definitive of a subsystem or subelement of the entire environmental management system — it is *nonspecific*. Under ISO 14001, however, a "program" is meant to be a "project" — it is *specific*. It is very important that you define this term with your auditors up front to avoid major confusion.

You can see the correlation between ISO 9001 and ISO 14001 listed in Table 8.1.

8.3 ISO 9001 REQUIREMENTS

Although ISO 14001 is much more explicit in requiring objectives, targets, and programs, there is enough information available in the ISO 9001 structure to allow the use of the quality structure to define your environmental requirements. Let's look at what the various ISO 9001 elements say:

Element 4.1.3 states: The supplier's management with executive responsibility shall review the quality system at defined intervals sufficient to ensure its continuing suitability and effectiveness in satisfying the requirements of this International Standard and the supplier's stated quality policy and objectives ..."

Element 4.1.1 states: The supplier's management with executive responsibility shall define and document its policy for quality, including objectives for quality and its commitment to quality..."

Element 4.2.3(b) states: The supplier shall give consideration to the following activities, as appropriate, in meeting the specified requirements for products, projects or contracts: (a) the preparation of quality plans; (b) the identification and acquisition of any controls, processes, equipment (including inspection and test equipment), fixtures, resources and skills that may be needed to achieve the required quality; ..."

Element 4.14.3 states: The procedures for preventive action shall include: ... (b) determination of the steps needed to deal with any problems requiring preventive action; (c) initiation of preventive action and application of controls to ensure that it is effective; ... "

8.4   SCENARIO

For the rest of this section, I am going to use a manufacturing process scenario as an approach to show how the environmental and quality requirements can be integrated. The manufacturing process will have quality problems, environmental aspects, and environmental legal concerns. The intent is to show the process of identifying an objective and a target and then establishing a project team that has the responsibility for solving the issues being presented. The project team will have to identify specific action items to be taken, when they will be initiated, and responsibility for completing those corrective actions. First, let's look at the scenario:

> XYZ Co. manufactures a widget used in the automotive industry as part of an electronic sensor system. Due to tight automotive specifications and difficulties in manufacturing, XYZ Co. is only yielding 50% through final inspection. Some of the process loss comes from a metal part used in the widget manufactured by a subcontractor. This part is made using a degreasing operation utilizing an ozone-depleting solvent. At XYZ Co. they manufacture the widget using various volatile organic chemicals that are vented right to the atmosphere and a cleaning operation using a slightly corrosive cleaning solution. The process also generates a liquid hazardous waste, a solid hazardous waste, and solid nonhazardous waste.

Let's list the various quality and environmental aspects in Table 8.2. So what we have now as a management team are six aspects which must be addressed in order to improve the yield (i.e., minimize the impact on quality) and reduce other waste (i.e., minimize the impact on the environment). As you can see from the table, some of the objectives have more than one target and, as a result, may have a very diverse project list (shown by project numbers). I have deliberately left out the details of the project list and, instead, have been presented in Table 8.3 with the details of the various projects.

8.5   WHAT AUDITORS WILL LOOK FOR

It is important to understand all of detailed requirements listed in ISO 14001 Element 4.3.3 and ensure that they are considered in fulfilling the requirements for Element 4.3.4. As stated in Element 4.3.3, the identification of your objectives and targets must take into consideration your significant environmental aspects, any applicable legal requirements, all available technological options, and financial implications. Your project list must show evidence that these have been considered. Additionally, when reviewing and determining what steps have been taken to achieve the targets, a veteran auditor may be able to assess whether or not all appropriate groups have been involved.

As an example, if we consider a project to reduce the amount of hydraulic oil waste in a pressing operation, then the project team should include maintenance and the project list include an evaluation of the preventive maintenance schedule for the presses. An action item may be to evaluate the preventive maintenance schedule for replacing the oil — an analysis of the oil for degradation may show that the replacement every six months can be extended to every 12 months (i.e., a 50% reduction).

TABLE 8.2
Objectives and Targets for Widget Production

#	Aspect	Impact	Objective	Target	Project #	Project Leaders
1	Production yields	Financial losses	Improve yields	(a) Improve to 75% by (date)	97-001	Process Engineer QA Manager
2	Subcontractor-made component	Increased scrap due to poor quality	Evaluate quality	(a) Improve to 90% by (date)	97-002	Purchasing Manager Process Engineer QA Manager
3	Subcontractor uses an ODS	ODS phase out; Labeling requirements	Eliminate ODS	(a) Eliminate need to clean, or (b) Replace with non-ODS solvent by (date)	97-003 97-004	Process Engineer Environmental Manager Product Technical Manager
4	Manufactured with a VOC	Air emissions	Eliminate VOC air emissions	(a) Change process, or (b) Use water-based solvent by (date)	97-005 97-006	Process Engineer Environmental Manager Product Technical Manager
5	Manufactured with corrosive cleaner	Toxic liquid waste	Eliminate corrosive cleaner	(a) Replace cleaner with noncorrosive by (date) (b) Eliminate cleaning by (date)	97-007 97-008	Process Engineer Environmental Manager
6	Process generates three solid waste streams	(1) Liquid hazwaste; (2) Solid hazwaste; (3) Solid non-hazwaste	(1) Reduce liquid waste; (2) Reduce solid waste; (3) Recycle	(1) Reduce 25% by (date) (2) Reduce 10% by (date) (3) Recycle 25% by (date)	97-009 97-010 97-011	Process Engineer Environmental Manager

TABLE 8.3
Project List for Widgets

Proj. #	Action Item(s)	Project Members	By When
97-001	(a) Evaluate quality of operating and production supplies and materials (b) Train all personnel on quality and statistical process control (SPC) techniques (c) Organize and train self-managed work teams (d) Evaluate equipment to identify primary issues causing downtime (e) Establish preventive maintenance program	Material Control Maintenance Personnel Quality Assurance Training Department Process Engineer	
97-002	(a) Define required specifications for part (b) Work with subcontractor to improve their process line to meet specifications (c) Identify alternative supplier of part	Quality Assurance Material Control Process Engineer	
97-003	(a) Work with subcontractor improve their process line to eliminate cleaning step with ODS	Process Engineer Material Control Environmental Manager	
97-004	(a) Evaluate non-ODS chemicals (b) Evaluate ODS chemicals which are scheduled for future phase-out (c) Determine changes in "pass through" labeling requirements (d) Determine other potential regulatory requirements for international marketing	Process Engineer Environmental Manager Shipping Department Product Technical Manager	
97-005	(a) Evaluate process to see if alternative manufacturing methods can be employed	Process Engineer	
97-006	(a) Evaluate water-based solvent (b) Qualify change in process per customer contract requirements	Process Engineer Environmental Manager Product Manager	
97-007	(a) Evaluate other cleaners which are noncorrosive (b) Work with Project Team #97-006 (joint effort)	Process Engineer Environmental Manager	
97-008	(a) Evaluate potential process changes to eliminate cleaning process	Process Engineer	

TABLE 8.3 (continued)
Project List for Widgets

Proj. #	Action Item(s)	Project Members	By When
97-009	(a) Evaluate potential process changes to eliminate liquid hazardous waste (b) Establish reduction targets and communicate on a regular basis with personnel (c) Work with Project Team #97-006 (joint effort)	Environmental Manager Process Engineer Department Management Operator work team	
97-010	(a) Evaluate potential process changes to eliminate solid hazardous waste (b) Establish reduction targets and communicate on a regular basis with personnel (c) Work with Project Team #97-006	Environmental Manager Process Engineer Department Management Operator work team	
97-011	(a) Evaluate sources of solid waste with outside analyst (b) Work with recycling vendor establish recycling center(s) (c) Train personnel	Environmental Manager Operator work team Solid Waste Analyst Recycling Vendor	

8.6 CONCLUSION

All of the sub-elements under the Planning requirements of ISO 14001 provide the structure for the continuous improvement process defined in your policy and, in my opinion, are the foundation of your whole program — if you don't do this section well, the rest of the program cannot take shape and becomes a moot point as far as auditing is concerned.

I hope that the information presented in this chapter will provide a good starting point. The major key to successfully integrating these requirements into ISO 9001 lies in the Design Review program. Defining the requirements for ISO 14001 into your Design Review and Purchasing/Contract/Material Control documentation can help streamline the process — but don't forget to write a *procedure* for defining the legal and other requirements.

Once you have made it through 4.3.3 and 4.3.4, the use of tables to summarize your information is much simpler and provides an at-a-glance view for the rest of the planning process. All one has to do is maintain a file for each project number that consolidates all the information (meeting minutes, reports, analyses, etc.) needed for an auditor to review.

Part IV

Implementation and Operation

9 Structure and Responsibility

9.1 INTRODUCTION

In the previous chapter, we focused on the elements that provide the primary plan behind the continual improvement process and, ultimately, the goal of prevention of pollution. In this chapter, we will focus on those elements of ISO 14001 that develop the capabilities and support mechanisms needed to achieve the objectives. The EMS guideline document, ISO 14004, defines it as identifying

> the capabilities and support required by the organization ... in response to the changing requirements of interested parties, a dynamic business environment, and the process of continual improvement. To achieve its environmental objectives an organization should focus and align its people, systems, strategy, resources, and structure.

If an environmental management system is to work, management must ensure that the organization is capable of meeting the requirements of the standards. This means management must:

- commit resources.
- integrate an EMS with other management systems.
- define responsibilities and accountability.
- build environmental awareness and motivation.
- identify the knowledge and skills needed.
- establish processes for communication and reporting.
- Implement operational controls.

For those of you already familiar with ISO 9001, you will have already noticed those same requirements are found in the quality standards. In the balance of this book you will find the integration process to be much more straightforward than that already discussed in the previous two chapters.

Element 4.4, *Implementation and Operation*, has seven sub-elements of which all but one (i.e., Subelement 4.4.7, *Emergency Preparedness and Response*) can be easily integrated into your quality system. Thus, the primary focus will be discussing the other six subelements, and a minor discussion of Subelement 4.4.7 will done for the sake of continuity.

I have found through my own efforts to integrate the two standards that the communication and conversation between myself, any internal auditor, management, and an employee who is being "interviewed" continues to focus back onto the specific procedures which give the employee direction on their job. These procedures have proven to be one of the best vehicles for providing employees an awareness

TABLE 9.1
Correlation of "Structure and Responsibility" Requirements

Requirement	ISO 9001	ISO 14001
Responsibility and Authority	4.2.2.1	4.4.1 (first paragraph)
Resources	4.2.2.2	4.4.1 (second paragraph)
Management Representative	4.2.2.3	4.4.1 (third paragraph)

of their responsibilities outside of a specific training session. The use of a Standard Operating Procedure (SOP), Maintenance Operating Procedure (MOP), Quality Assurance Procedure (QAP), etc. can be an effective vehicle for integrating ISO 9001 and ISO 14001. These types of documents will play a major part throughout this chapter.

9.2　STRUCTURE AND RESPONSIBILITY

Whether it is an environmental management system or quality management system, the very first thing an organization must do is to define its structure, responsibilities, and accountabilities. It is, foremost of all, critical for the senior manager to set the "tone" of the organization's objectives by first communicating his or her commitments to the establishment of a management standard. Without that commitment, any program or system will be doomed to failure and end up being the "flavor of the month" — a great deal of energy, resources, time, and finances will be expended needlessly. The success or failure is entirely in the hands of senior management.

For much of the process discussed in this book, most organizations will already have specific procedures in their QMS-related system to allow the integration to occur. In integrating the requirements for this section, however, it is best to look at your quality manual as a starting point rather than a specific procedure. Although a detailed evaluation of the quality and environmental manuals will be done later in Chapter 12, we will briefly refer to the manuals here for the purpose of seeing how the integration for Structure and Responsibility can occur. In sections 9.2.3 through 9.2.5, I will not be spending too much time discussing the issues in detail, but will provide some preliminary information on what the standards require. Section 9.2.6 will provide the wrap-up discussion.

9.2.1　THE RELATIONSHIP BETWEEN ISO 9001 AND ISO 14001

If you place the requirements for both of the standards in this area side-by-side, you will immediately notice almost a word-for-word similarity. The drafters of ISO 14001 used the wording in ISO 9001 as the template for Element 4.4. The correlation of the two standards can be seen in Table 9.1.

9.2.2　DEFINING THE STRUCTURE

One of the easiest ways for an organization to define their structure is through an organizational chart — it is the best way to present a "snapshot" of authority. Many organizations typically show an organizational chart such as that shown in Figure 1.

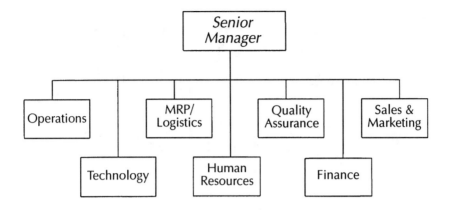

FIGURE 1. Typical senior management structure.

The problem with this chart is that it does not give the viewer or an auditor a true perception of how the organization really functions and, of course, for the sake of this discussion, does not reflect the requirements of ISO 14001. Your manual can show much more detailed organizational structure such as that shown in Figure 2. This figure shows the Environmental Manager directly reporting to the organization's senior manager. In many cases, however, the Environmental Manager is typically located under the Operations Manager, since this is where most of the environmental aspects and impacts occur. In either case, it is very important to have the Environmental Manager report as high up in the organization as possible — the importance of developing and implementing an EMS cannot be buried too far down in an organization, but it must be visibly evident that senior management gives it high priority. For those organizations which choose to have the Environmental Manager report lower down, it is very easy to expand the organizational chart in your manual to reflect all levels of the organization including the Environmental Manager.

9.2.3 RESPONSIBILITY AND AUTHORITY

It is critical to any program that each and every person understand their responsibilities and the extent of their assigned authority within those responsibilities. ISO 14004 offers a list of issues that can be considered, and a few of these are listed here:

- What is the relationship between environmental responsibility and individual performance and is this periodically reviewed?
- How do the responsible and accountable personnel:

  — understand the consequences of noncompliance?
  — understand the accountability which applies to them?
  — anticipate, identify and record any environmental problems?
  — initiate, recommend, or provide solutions to those problems?
  — initiate action to ensure compliance with environmental policy?

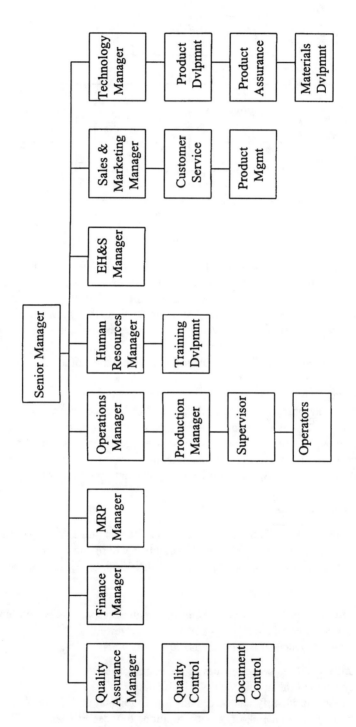

FIGURE 2.   Expanded organizational structure.

You will notice that this list is primarily focused on ensuring compliance with the ISO 14001 EMS — this can be a starting point to which the quality requirements can be added. If you look at the ISO 14001 requirements you will note that they are very simple and direct:

Element 4.4.1 (first paragraph) states: Roles, responsibility and authorities shall be defined, documented, and communicated in order to facilitate effective environmental management.

Whereas the requirements under ISO 9001, **Element 4.2.2.1**, are much more extensive than ISO 14001, as can be seen below:

> The responsibility, authority, and the interrelation of personnel who manage, perform and verify work affecting quality shall be defined and documented, particularly for personnel who need the organizational freedom and authority to: (a) initiate action to prevent the occurrence of any nonconformances relating to the product, process and quality system; (b) identify and record any problems relating to the product, process and quality system; (c) initiate, recommend or provide solutions through designated channels; (d) verify the implementation of solutions; and (e) control further processing, delivery or installation of nonconforming product until the deficiency or unsatisfactory condition has been corrected.

9.2.4   RESOURCES

Resources can mean different things to different managers, but is generally understood to be people, time, and finances.

The requirements in ISO 9001 state:

ISO 9001, Element 4.2.2.2 states: The supplier shall identify resource requirements and provide adequate resources, including the assignment of trained personnel, for management, performance of work and verification activities including internal quality audits.

The ISO 14001 standards are very similar and more specific in naming some of those resources:

Element 4.4.1 (second paragraph) states: Management shall provide resources essential to the implementation and control of the environmental management system. Resources include human resources and specialized skills, technology and financial resources.

9.2.5   MANAGEMENT REPRESENTATIVE

The designation of a management representative will be dependent on the size of the organization, the level of management commitment, and/or the ability financially to maintain an "expert" on the payroll. For many small-to-medium sized companies (SMEs) this last point may be of primary concern. Thus, an owner may designate his/herself as the representative or designate an employee with an engineering or technical background which may or may not be specifically related to either the quality or environmental fields. If this is the case, then management must take steps to ensure that the requirements of both ISO 9001 and ISO 14001 are met through education or by employing an outside consulting firm to assist in establishing the program requirements. The use of an outside consulting firm, of course, may still place a financial burden on a smaller business. The need for a senior manager to become "educated" about the quality and/or environmental arena (whether another "expert" is available) is not necessarily a waste of time since he/she will be able to provide better guidance for the organization as a whole and be a much better driving force for the implementation.

Let's now take a look at the individual ISO requirements that you will note are almost identical:

ISO 9001, Element 4.1.2.3, states: The supplier's management with executive responsibility shall appoint a member of the supplier's management who, irrespective of other responsibilities, shall have defined authority for (a) ensuring that a quality system is established, implemented and maintained in accordance with this International Standard, and (b) reporting on the performance of the quality system to the supplier's management for review and as a basis for improvement of the quality system.

ISO 14001, Element 4.4.1 (third paragraph) states: The organization's top management shall appoint a specific management representative(s) who, irrespective of other responsibilities, shall have defined roles, responsibilities and authority for (a) ensuring that environmental management system requirements are established, implemented and maintained in accordance with this International Standard, and (b) reporting on the performance of the environmental management system to top management for review and as a basis for improvement of the environmental management system.

The key combined requirements of the representative are to: (a) ensure that system requirements are *established*, *implemented*, and *maintained* in accordance with the International Standards; and (b) to *report* on the performance of the system to top management for review.

9.2.6 COMBINED JOB DESCRIPTIONS

So, with these requirements in mind, let's write job descriptions for various levels of management that reflect both ISO 9001 and ISO 14001...
Senior Manager   has the ultimate responsibility for:

- the quality of all products and services provided, *as well as the overall responsibility for the environmental performance* of the organization.
- defining the organization and providing the necessary resources to achieve the organization's mission, *including the appointment of an Environmental Manager.*
- ensuring that all activities affecting Quality *and the Environment* are conducted in a planned and systematic fashion.
- the Quality Improvement System (QIS), the Quality Assurance Standards, *and the Environmental Management Standards* as set forth in this manual, are being managed.

Quality Assurance Manager   is the organization's Management Representative for all quality matters and has the authority and responsibility for:

- ensuring that the policies, practices, and procedures, as set forth in this manual, are understood, implemented, maintained, and continuously improved.
- structuring the Quality Improvement and Quality Assurance Systems to best ensure compliance with all quality requirements.
- planning and coordinating management reviews of both the Quality Improvement System and Quality Assurance System and Standards in order to facilitate maintenance and continuous improvement.
- supporting the efforts of the *Environmental Manager* in preparation of, and during, internal and external audits.
- ensuring environmental reviews are part of the Design Review Program.

Sales and Marketing Manager   Primary responsibilities are to:

- direct the sales, marketing, and product management organizations to ensure integration while achieving profitable growth.
- direct programs to both measure and improve customer satisfaction.
- direct processes to ensure that customer requirements for quality, delivery, and performance are clearly understood and can be acted on, *including customer requests concerning product safety and environmental impacts* — this includes analyzing, interpreting, and translating customer requirements into technical product specifications.
- direct efforts to research, quantify, and characterize customer needs, *including a product's environmental requirements and potential environmental benefits*, then develop worldwide marketing and product line strategies that will satisfy those needs.
- lead the effort to find new applications of product.
- direct the process that enters, administers, and coordinates the fulfillment of customers' orders.
- direct the development and continuous improvement of procedures to create customer awareness, to provide training for customers as well as the sales force, and to make it easier for customers to deal with the organization.

Operations Manager   Primary responsibilities are to:

- manage, maintain, and continuously improve the organization and processes of the organization to manufacture, document, test, package, and deliver products in full compliance with customers' expectations *and to minimize or eliminate any environmental impacts*.
- support the sales, marketing, product management, technical, and logistics organizations to ensure that successful implementation of commitments is feasible.

- direct the manufacturing, documentation, packaging, controlling, and delivery of products that meet the organization's specifications and customers' stated and explicit requirements.
- implement the use of statistical methods to effectively control and continuously improve processes.
- support training programs as a means to ensure personnel are operating safely, are *in compliance with any job-related regulatory requirements* and are *minimizing the impact of their job on the environment.*
- ensure systems are always open to facilitate communication with employees.

Technology Manager   Primary responsibilities are to:

- lead the development and maintenance of technical strategy, including a patent strategy and portfolio.
- lead the development of new products as well as new process methodologies in order to improve the quality, lower the cost of product, and *minimize the environmental impact of new/modified products.*
- lead the development of advanced materials and process technologies in order to help the organization better serve its customers' needs.
- direct the development of advanced test and measurement technologies and supporting equipment to qualify the performance of the product.
- support the implementation and philosophy of Product Stewardship and Design for Environment (DFE) through the Design Review Program.
- support the senior management team and environmental management in assessing environmental abatement control technology as a means to minimize the organization's environmental impact.

Manufacturing Resource Planning and Logistics Manager   Primary responsibilities are to:

- direct the design, development, and implementation MRP, including an overall process for Sales and Operations Planning, Master Production Scheduling, Material Requirements Planning, and Production Activity Control.
- develop, track, and report metrics for assessing progress and effectiveness of implementation.
- design, develop, and implement MRP training.
- drive the continuous improvement program for on-time delivery.
- develop processes for managing customer complaints and returns.
- work with Environmental Management to minimize or eliminate the purchase and use of hazardous materials.

Finance Manager   Responsibilities are to:

- direct accounting, financial planning, and analysis.

- direct the establishment and continuous improvement of financial information systems.
- ensure product cost documentation is accurate for management decision-making and accounting.

Human Resources Manager   Primary responsibilities are to:

- work with management to identify the skills, abilities, knowledge, and types of people needed for the organization.
- facilitate the design and implementation of organization-wide personnel training, development, and succession process.
- design and facilitate the implementation of an appropriate organizational structure, as well as change management and team development processes.
- provide internal consultation for the development of individuals and teams as a means to ensure processes are being operated safely and producing minimal impacts on the environment.

Production Manager   Primary responsibilities are to:

- provide support to supervisors immediately under their direction.
- budget needed finances.
- provide resources to ensure customers' expectations are being met as they relate to product quality and delivery requirements.
- provide engineering support to ensure processes are safe for personnel to operate and *for minimizing its potential impact on the environment*;

Supervisors   As the "front-line" manager, supervisors have the primary responsibility for ensuring the day-to-day management of the quality and environmental programs and adherence to the organization's operational policy. This includes the following:

- designating area representatives for quality and environmental emergency situations and allowing time for training and meetings.
- administering the disciplinary policy in the event of employee noncompliance.
- participating in any internal quality and/or environmental audits and ensuring required corrective actions for noncompliance are completed in the timeframe(s) specified.
- completing quality and environmental inspections as scheduled and implementing and completing any corrective actions in the timeframe(s) specified.
- scheduling department-specific meetings as required to discuss pertinent quality and environmental issues.
- ensuring all employees under their direct management attend all required training as specified under the training needs analysis.

- new hire orientations are conducted before work begins and to ensure the orientations include all relevant quality and environmental information and to inform them of the impact their job has on quality and the environment.
- proper job instruction is given to ensure employees understand the impact their job has on quality and the environment.
- ensuring all quality and environmental data is maintained as specified by the program and/or legal requirements.

Last, but not least:

Environmental Manager   has the primary responsibility for:

- reporting to senior management on the status of the environmental management program and ensuring that the EMS requirements are established, implemented, and maintained in accordance with ISO 14001.
- providing leadership in ensuring full compliance with any applicable legal and other code/standard requirements.
- ensuring compliance with internal and third party auditing requirements.
- providing direction, resources, and guidance on any training for employees.
- maintaining all documentation in accordance with the requirements set forth under the ISO 14001 EMS and all other legal and/or regulatory requirements.
- working with the *Quality Assurance Manager* to integrate the quality and environmental systems.
- ensuring monitoring, measurement, and reporting requirements in achieving the environmental objectives and targets are established, maintained, and reported.
- providing input into the environmental requirements of a Design Review.

The job descriptions presented above for the Quality Manager and the Environmental Manager are indicative of job descriptions intended for representatives whose sole responsibility is either quality or environmental, respectively. In the event a representative (let's call him/her the Management Systems Manager for example) is appointed with responsibility for both, it would be very easy to combine the requirements under both standards in the following manner:

Management Systems Manager   shall have the responsibility to:

- ensure the quality and environmental systems are established, implemented, and maintained in accordance with the International Standards.
- report on the performance of the quality and environmental systems to management for review and as a basis for improvement of the quality and environmental management systems.

And so on ...

9.2.7   Nonmanagement Personnel

As mentioned in the introduction of this chapter, a Standard Operating Procedure can be an excellent source for defining the roles, responsibilities, and authorities of nonmanagement personnel, especially in the area of operations. A well-written procedure for a job function can define the responsibilities for operating personnel, maintenance, technical, and engineering support.

I'm assuming in most cases that the majority of you reading this book have gone through the ISO 9001 process and have potentially been involved in the development of procedures for various document levels and a variety of job descriptions. If this is the case, you are already one step ahead and, as a result, can very easily integrate the requirements for ISO 14001.

Before we proceed too much further, I would highly recommend you evaluate the various sections of your procedures and consider whether or not they are adequate. It is my opinion that a well-structured procedure at a minimum should have the following sections:

- Purpose
- Scope
- Definition of terms
- References
- EH&S precautions
- Precedence
- Responsibilities
- Materials
- Tools
- Procedure

It would be expected that a document describing the functions of a piece of machinery would have all of these sections, whereas higher level documents would not necessarily contain "Materials" and "Tools." Additionally, higher level documents may have little or no "EH&S Precautions" and less-defined "Procedures." I am not suggesting you go back and rewrite every one of your procedures, but if your current sections do not have something similar or a logical place to put the ISO 14001 environmental requirements, it might behoove your organization to consider a document revision before proceeding with the integration.

The above list of sections are from the document system I have worked with for the past several years. What I found as I went through the integration process was that the system in place would readily accept the requirements of ISO 14001 without having to make any overall document changes (i.e., changing the procedure for writing a procedure). For the purposes of this section, the "Responsibilities" section is the obvious choice.

Below is a list of potential responsibilities for the personnel involved in a process which uses chemicals:

- Ensure the Standard Operating Procedure (SOP) is followed.
- Ensure product integrity, quality, and traceability.
- Understand any statistical tools needed to produce a quality product.
- The Process Engineer/Technician is responsible for developing, implementing, and updating SOP.
- Follow and adhere to all environmental, health, and safety precautions listed.
- Ensure proper solvent is being used.
- Understand the consequences of not adhering to the procedure and the potential impact that failure may have on the product, yourself and your fellow employees, and the environment.
- Wear appropriate personal protective equipment.
- Report any accidental chemical releases, especially if released to a sewer or stormwater source.
- Understand any legal or other standards/codes which may have an impact on your job.
- Understand the emergency shutdown procedures.
- Ensure proper disposal of any hazardous waste.
- Understand the requirements for the recycling of materials.

You may notice that some of these responsibilities also overlap with the "awareness" requirements in Section 10.4, but you will find that ISO 14001's intention for the "awareness" requirements goes much further and is more definitive.

9.3   WHAT AUDITORS WILL LOOK FOR

In defining the role, responsibilities, and authority for the Environmental representative (Section 9.2.6) under the specific ISO 14001 requirements, an auditor will look for two key descriptions that are listed as points "a" and "b" under Element 4.4.1. You can very easily avoid a negative audit finding by virtually copying these two paragraphs directly into the job description. If you have not already noticed, the job description directly above contains these two requirements in the first bullet item! In the event the "representative" has some combination of responsibilities, an auditor will look to ensure that the listed responsibilities for both of the standards are combined into that person's job description. An example of this has already been presented.

Another critical item to watch out for in an integrated system concerns the demonstration or evidence of "resources" being available and a specific written requirement for management to be held responsible for providing those resources. Both of the standards provide a list of potential resources and it is highly recommended that management give due consideration for any others.

10 Training, Awareness, and Competence

10.1 INTRODUCTION

We now start what I consider to be one of the most critical segments of both ISO 9001 and ISO 14001 in terms of its impact on the success of the system. In addition to an effective internal communications program, the training, awareness, and competence of an organization's employees can determine the "tone" of your entire program (e.g., excellent, good, fair, or poor). The finest and best documented management system in the world is only as good as the level of an organization's employee training, awareness, and competence and vice versa. An excellent management system takes an excellent documented program and excellent employees to put everything into place and activate it.

Table 10.1 outlines the sections of ISO 9001 and ISO 14001 that bear a relationship in the training requirements.

10.2 A COMPARISON OF THE STANDARDS

As you can see from the table, ISO 9001 does not have any specific mention of "awareness," but it is obviously implied throughout the standards. Specifically, ISO 9001, Section 4.18 states:

> The supplier shall establish and maintain documented procedures for identifying training needs and provide for the training of all personnel performing activities affecting quality. Personnel performing specific assigned tasks shall be qualified on the basis of appropriate education, training, and/or experience, as required. Appropriate records of training shall be maintained.

In contrast to ISO 9001, the ISO 14001 Environmental Management Standards have much more stringent requirements. This may be partly due to the potential for greater consequences or impacts on the environment in comparison to product quality. Let's now look at what ISO 14001 says about training:

ISO 14001, Element 4.4.2, states:

> The organization shall identify training needs. It shall require that all personnel whose work may create a significant impact upon the environment, have received appropriate training.

> It shall establish and maintain procedures to make its employees or members at each relevant function and level aware of (a) the importance of conformance with the environmental policy and procedures and with the requirements of the environmental

TABLE 10.1
Correlation of "Training" Requirements

Requirement	ISO 9001	ISO 14001
Training Needs and	4.18	4.4.2 (first paragraph)
Appropriate Training Awareness		4.4.2 (second paragraph — parts a–d)
Competence/Qualified	4.18	4.4.2 (third paragraph)

management system; (b) the significant environmental impacts, actual or potential, of their work activities and the environmental benefits of improved personal performance; (c) their roles and responsibilities in achieving conformance with the environmental policy and procedures and with the requirements of the environmental management system, including emergency preparedness and response requirements; and (d) the potential consequences of departure from specified operating procedures.

Personnel performing the task which can cause significant environmental impacts shall be competent on the basis of education, training, and/or experience.

10.3   TRAINING

Both of the standards require two specific things under training: (a) a training needs analysis; and (b) employees must receive training commensurate with their function. Executive and management personnel, for instance, may have training which provides them with an overview of the environmental and/or quality systems — training which will help them to make decisions and determine the effectiveness of the system(s). Supervisors and operating personnel in contrast must have more specific training in the procedures and skills required (i.e., documentation, the relationship of their jobs to the overall quality of the product, the actual and potential impacts of their job on the environment, etc.) One thing to keep in mind at all times, however, is that the training needs should take into account what the objectives are for the organization, whether it be quality or environmental. At the heart of the analysis is the need to analyze each specific job function as to their potential impacts on the environment and/or quality. Typical factors to be considered are:

- specific regulatory compliance issues pertinent to the job function.
- specific corporate/industry standards and codes compliance.
- data collection techniques.
- safety and health factors.
- special chemical handling requirements.
- understanding special analysis or measurement techniques.

Larger organizations typically have training departments and training coordinators, but the majority of the small to medium-sized companies do not have this luxury. In all cases, however, a common requirement is to have qualified individuals assess each job function and make a determination as to what the training requirements will be. Organizations may find that there is a set of core training requirements for all employees and small list of specific training needs determined by the job function. Core training may include some of the following:

- New hire orientation
- EH&S management systems, including policy
- Quality management systems, including policy
- Emergency preparedness and response
- Computer software systems
- Material management systems
- Presentation skills
- Teambuilding
- Managing effective meetings
- Statistical Process Control (SPC)
- ISO 9001 Standard
- ISO 14001 Standard

Selected or specific training may include the following:

- Design for Manufacturability (DFM)
- Failure Mode/Effects Analysis (FMEA)
- Design for Environment (DFE)
- Design review process
- ISO internal auditing techniques
- Taguchi methodology
- Gage R&R
- Project management
- Chemical hazard communication
- Environmental regulations
- Financial analysis
- Chemical spill response
- Accident/incident investigation
- Hazardous waste management

As an example, let's analyze and list potential training needs for a Process Engineer of a manufacturing line that requires the use of a chemical to manufacture a product. The reason I have chosen this profession is because a Process Engineer typically requires a very extensive list of specific training needs. Table 10.2 shows such a list of both core and specific training needs for the Process Engineer. It includes the approximate hours required to conduct the training and who is responsible for giving the training.

Each and every job description within an organization can have a listing such as in Table 10.2 which easily details all training required in the areas of quality, environmental, health and safety, engineering, human resources, information technology, technical/developmental, and any other specialized skills. One procedure can be written which consolidates all of the training requirements from ISO 9001 and ISO 14001.

TABLE 10.2
Training Needs for Process Engineer

Core Training	Length (h)	Source
New Hire Orientation	4	Human Resources
EH&S Management Systems	2	EH&S
Quality Management Systems	4	QA/QC
Statistical Process Control (SPC)	16	Engineering
Computer Software Systems (selected)	24	Information Technology
Material Management Systems	16	Purchasing/Logistics
Shop Floor Control	3	Engineering
ISO 9001 Standard	4	QA/QC
ISO 14001 Standard	4	EH&S
Managing Effective Meetings	8	Human Resources
Teambuilding	8	Human Resources
Emergency Preparedness and Response	1	EH&S
Specific Training		
Design for Manufacturability	16	Engineering/Development
Failure Mode/Effects Analysis	16	Engineering/Development
Design for Environment	8	EH&S/Development
Taguchi Methodology	8	QA/QC
Chemical Hazard Communication	2	EH&S
Project Management	4	Engineering/Development

10.4 AWARENESS

The next requirement in the standards concerns whether or not your employees are aware or cognizant of the consequences of failing to follow procedures. Again, the consequences can impact both quality and the environment and, although ISO 9001 does not have any direct reference to "awareness," it obviously is a factor which must be taken into a great deal of consideration.

As mentioned in Section 9.2.7, a Standard Operating Procedure (SOP) is the most logical place to inform employees of what the "awareness" requirements are. To accomplish this, I will again use examples of manufacturing processes that utilize chemicals generating both hazardous and nonhazardous wastes and air emissions. For the sake of quality, of course, be aware that the use of these chemicals may require certain specifications of volume, application characteristics, purity, and other requirements such as Military Specifications.

As done throughout previous sections, I will provide a detailed list of examples which will provide an employee an "awareness" of his/her responsibilities and the potential or actual impacts his/her job function may have on quality and the environment:

- It is the responsibility of the operator to maintain chemical purity.
- It is the responsibility of the operator to make sure all temperature controllers are in calibration.
- It is the responsibility of the operator to understand the operating procedure exactly as it is written.
- It is the responsibility of the operator to understand shutdown procedures in the event of an emergency. *Failure to understand the procedure could result in either an uncontrolled spill and/or fugitive air emissions.*
- Wear chemical splash goggles when operating or maintaining the equipment. *Failure to do so could result in the chemical splashing into an eye.*
- This process requires a Permit-to-Operate from the Air Resources Board (ARB). Permit conditions require a log to be maintained showing the flux usage over a 12-month consecutive cycle. *Failure to log flux usage could potentially result in a citation and fine, and, in the worst extreme revocation of the permit and shutdown of the operation.*
- Rags and other wipes used to clean equipment are classified as hazardous waste and must be disposed of in the proper container. The container must be properly labeled and the lid sealed each time waste is placed into it. *The container must be removed within 90 days from the accumulation start date noted on the hazardous waste label or a possible fine could be levied by the EPA.*
- The chemical is classified as a hazardous waste and must be placed in a plastic, 5-gallon, closed top container. *The hazardous waste label must be labeled according to regulatory requirements — all sections of the label must be completed or a possible fine could be levied by the EPA or DOT.*
- Any chemical residue that ends up on the floor must be placed into an appropriate container for final disposal into the trash compactor. *Residue must be cleaned up immediately and is not to be swept or washed into the sewer/drain system. This would violate water discharge permit requirements.*

As you can see I have highlighted the "consequence" statements in the above examples. Some of these are potential and some are actual consequences.

10.5 COMPETENCE

Competence is defined as "having requisite or appropriate abilities, knowledge, or skills" to complete a task or function. ISO 9001 defines it as being "*qualified.*" For both ISO 9001 and ISO 14001, this has a direct correlation to not only training in general, but also to the need to *develop* an individual employee. As discussed in Section 9.2, a training needs analysis must be developed for each employee so that they can become competent to do their job.

Competency can be accomplished and/or understood in four ways:

- On-the-job training
- Certifications and education
- Outside training courses
- Previous experience

In the case of previous experience, management would need to be provided evidence from a prior employer and the employee most likely would be able to verbally communicate his/her understanding of a process or equipment. For outside training courses, most programs provide certificates of completion and/or the results of some form of examination to demonstrate an understanding of the material.

For the rest of the discussion of competency, however, I would like to focus on the area of on-the-job training and provide examples of how the competency requirements can be satisfied. Specifically, the development of a certification process will be the foundation for ensuring competency and, of course, the most logical starting place is with a Standard Operating Procedure.

Although actual hands-on experience cannot be replaced, the ability of management to ensure an employee truly understands his job function can best be accomplished through the administering of a certification or examination program. A quiz or examination provides management with an opportunity to see if an employee understands the written procedure, the consequences of not following the procedure, and can test an employee on potential or actual troubleshooting capabilities. Although what follows may be obvious in light of the discussions and information given in the previous sections, I am going to provide sample questions on an examination for a manufacturing process which uses chemicals — however, I will not provide questions in the area of quality. The following types of questions can easily be integrated into a quality examination. You will note that several questions relate to Section 10.4 above. After each question, I also provide possible answers.

Q. When dispensing a liquid solvent into the delivery reservoir, what must be done immediately after?

A. Log the amount of solvent on the dispensing log in order to comply with the air permit conditions.

Q. What are some *potential* impacts on the environment this process may create?

A. (a) Failure to seal the delivery reservoir could cause liquid to spill onto the floor and enter a sewer drain line; (b) failure to replace the cap on the solvent container could cause fugitive air emissions; and (c) failure to replace the cap on the solvent container could cause a spill if the container is knocked over.

Q. If the solvent delivery system is not sealed and creates an increase in the consumption of the solvent, what is a potential impact in terms of regulatory compliance?

A. Since the solvent is classified as a hazardous air pollutant (HAP), this may potentially exceed the Title V regulations.

Q. What are some *actual* impacts on the environment this process creates?

A. (a) Liquid hazardous waste; (b) solid hazardous waste; (c) nonhazardous waste; and (d) air emissions.

Q. What is the environmental impact if the machine is not shut down over the weekend?

A. An increase in the consumption of *natural resources* such as electricity, natural gas, water, and other process gases.

Q. If preventive maintenance is not completed on the bearings for the compressor located on the roof, what is a potential environmental impact?

A. The *noise* level may increase causing complaints from the surrounding neighborhood.

Q. If polymer pellets are not swept up off the floor immediately after being spilled, what is a potential consequence?

A. The pellets may end up in the *sewer system* line.

Q. If you are taking trash outside to a dumpster that contains a nonhazardous powder, such as magnesium hydroxide, and you spill it, what should you do and why?

A. The chemical should be cleaned up immediately and not allowed to enter any *stormwater drain* — the drain flows into the wetlands along the bay.

Q. If you have some hazardous waste, into what type of container should it be put and what type of information should the hazardous waste label contain?

A. Five-gallon, closed top polyethylene drum and labeling requirements vary by area.

Q. Which materials do we recycle from the process?

A. (a) metal; (b) cardboard; (c) white paper; (d) solder dross; and (e) solvent through a recovery unit.

Q. In the event of a chemical spill, what should you do?

A. Alert personnel in the area and assist in placing containment pillows/socks around the spill to contain it. Contact the spill response team or the area supervisor.

Q. In the event a spill is greater than five gallons, what should you do?

A. (a) Contact site security for assistance; or (b) call 911; or (c) call spill response contractor posted by the telephone.

As you can see, the questions above address potential and actual impacts to the environment in the areas of air emissions, hazardous waste, nonhazardous waste, recycling, natural resource consumption, and community noise, as well as spill response activities and regulatory compliance.

Another means in which competency can be developed is through the activity associated with the internal auditing process. In order to verify the knowledge or competency of an individual, it is very easy to have a series of questions on an audit checklist which direct attention not only on an employee's understanding of his/her job function, but also on the consequences of their job on product quality and potential and actual environmental impacts. The results of the answers may cause the issuance of a corrective action, which in turn can create communication within a team meeting, a change in an examination, and/or a change in the procedure to further enhance the requirements. The internal communication requirements can be used by management to discuss the corrective actions stemming from the audit and, thereby, increase awareness and competence.

10.6   WHAT AUDITORS WILL LOOK FOR

As stated earlier on in this chapter, if you already have a training program in place, it is very easy to integrate your environmental training requirements. For ISO 14001, an auditor will focus on determining whether or not a training needs analysis has adequately addressed the communication of the overall EMS requirements (i.e., policy, roles and responsibilities, emergency preparedness). It should also address an understanding of their respective *significant* environmental aspects (actual and potential).

The guideline document, ISO 14004, offers some insights into an auditor's mind and you should consider them in developing your training program — with a slight addition for ISO 9001:

- How does the organization identify environmental (and quality) training needs?
- How are the training needs of specific job functions analyzed?
- Is training developed and reviewed and modified as needed?
- How is the training documented and tracked?
- Is the effectiveness of training evaluated periodically?

One last comment   Although this will be dealt with in a later chapter, you should be aware of the competence of contractors at the site. An unskilled contractor or a contracted employee who is not aware of his/her job requirement can have an actual or potential impact on the environment, be in noncompliance with a regulatory requirement, be violating a company policy, code, or standard, and/or impacting an environmental objective.

11 Communication

11.1 INTRODUCTION

The most successful quality or environmental program is founded upon the principle of teamwork. Despite the clear responsibilities of management in most cases, management cannot nor is it expected to take full responsibility. A successful program is only successful if *everyone* takes responsibility, and the most effective tool is a communications program.

Through personal (one-on-one) meetings, group meetings, written communiqués, general promotions, and customer service organizations, the goals and objectives of a quality and/or environmental program can be successful and create a spirit of partnership with employees, vendors, customers, and the community. However, a successful communications program can only be effective if it occurs from the "top down." Management's responsibility is to deal with the concerns and issues about quality and environmental questions of their business' activities, products, and/or services and to communicate on a daily basis their commitment to quality and environmental objectives and to the welfare of their stakeholders (i.e., employees, vendors, customers, stockholders, and the community) and the environment itself.

11.2 COMPARISON OF THE STANDARDS

Table 11.1 shows how ISO 14001 and ISO 9001 interrelate. As you can see, there are several gaps in the area of internal communication. What you will notice is that I have also included several other requirements for "communicating" various aspects of ISO 14001 and categorized them as either internal or external. Although the standards do not directly state these as communication requirements, it is specified that employees "understand" — this implies that some form of communication must take place.

11.3 INTERNAL COMMUNICATIONS

When talking about "internal" communications, we are not only referring to the types or *how* we want to communicate, but also *what* we are wanting to communicate — *what* we want to communicate can influence the *how*.

Before we define the *how* and the *what*, let's first define the purpose and the scope of internal communications whether it concerns quality or environmental — both require the communication of information.

TABLE 11.1
Correlation of "Communication" Requirements

Requirement	ISO 14001	ISO 9001
Internal Communication	4.4.3(a)	4.18
— Policy	4.2(e)/4.4.2(a)	4.1.1
— Significant Impacts	4.4.2(b)	
— Roles and Responsibilities	4.4.2(c)	4.1.2.1
— Emergency Preparedness	4.4.2(c)	
— Procedures	4.4.2(d)/4.4.6(a,b)	4.2.2/4.9
— Design Control		4.4.3/4.4.6
— Audits	4.5.4(b)	4.17
— Management Review	4.6	4.1.3/4.14.3(d)
External Communication	4.4.3(b)	4.14.2(a)
— Policy Publicly Available	4.2(f)	
— Contracts		4.3.2
— Suppliers/Contractors	4.4.6(c)	4.6.2
— Design Review		4.4
— Customers	4.4.3(b)	4.14.2

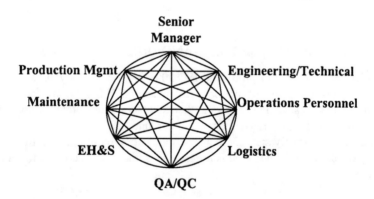

FIGURE 1. The "ideal" internal communications web.

11.3.1 PURPOSE OF INTERNAL COMMUNICATION

The overriding principle behind an internal communications system is to facilitate a smooth operation — the flow of information is just as important as the flow of product through a manufacturing plant. Defects and poor yields in manufacturing can cost a firm money — poor communication can do the same thing. As stated in the introduction of this section, it also helps create and should foster the development of teamwork and a sense of worth and contribution to the final goal.

11.3.2 SCOPE OF INTERNAL COMMUNICATION

Internal communication must be from the "top down" — from the highest manager within the organization down to the production worker who is at the heart of making the product. The communications network must allow each and every employee equal access to all levels of the organization and an "open door" policy must be obvious to each and every employee — a good manager never "shuts the door." In an ideal situation, the communications matrix (web) should be like that in Figure 1. This web not only shows that communication must cross all boundaries, but that listening allows the communication to occur.

11.3.3 CONTENTS OF INTERNAL COMMUNICATION

No matter whether the type of communication is to strictly inform, to lend a listening ear, or to seek consultation, a manager must take the primary responsibility for the content and the quality of the communication within their areas of responsibility. In terms of content and for the purposes of this book, a manager will need to educate his/her employees on the following ISO 14001 and ISO 9001 elements:

ISO 14001	ISO 9001
• Environmental Policy	• Quality Policy
• Legal and Other Requirements	• Compliance with reference standards, etc.
• Results of Management Reviews	• Management Reviews
• Results of Audits	• Results of Audits
• Results of Monitoring and Measuring Activities	• Monitoring and Control of Suitable Process Parameters
• Responsibilities of Employees	• Responsibilities of Employees
• Training Requirements	• Training Requirements
• Operational Controls	• Process Controls
• Corrective and Preventive Actions	• Corrective and Preventive Actions
• Environmental Objectives and Targets	• Quality Objectives and Targets
• Environmental Aspects	
• Emergency Response Procedures	

As you can see, there is a great deal of overlap in the area of communication. Meetings or any other form of communication can easily have an agenda that discusses both quality and environmental issues. For example, a combined internal audit report (to be discussed in the next chapter) can be the subject of a team review that results in both quality and environmental corrective actions.

11.3.4 TYPES OF INTERNAL COMMUNICATION

Table 11.2 contains a matrix that shows the various types of quality and environmental communication methods that can be used by management. In all of the

TABLE 11.2
Internal Communications Matrix

Method	Responsibility	Purpose	Audience	Content
Plant Meeting	Senior Manager	Overall business performance	All Employees	Revenues; sales; QA; EHS metrics; etc.
Business Reviews	Business Team Coordinators	Performance of market business	Mgmt and staff as appropriate	Status: business units; market strategy
CIT* Reviews	CIT Leaders	CIT performance	Mgmt.; staff; CIT members	Ongoing quality and environmental issues
Department or Shift Meetings	Line Mgmt./Supervisors	Local department; business updates	All dept. or shift members	QA; EHS; problem-solving; etc.
Staff Meetings	Senior Managers	Future issues; updates	Senior mgmt. staff	Current and future business issues
CAT** Reviews	CAT Leaders	CAT performance	Mgmt.; staff; CAT members	Specific corrective action activity
Newsletter	EHS & QA Managers	General info; QA/EHS status or metrics	All employees	Overall or specific business, QA, or EHS information
Bulletin Board	Line Managers/Supervisors	General info; QA/EHS status or metrics	All employees	Overall or specific business, QA, or EHS information
Procedures	Line Mgmt./Supervisors	Specific job or task activities	All employees as appropriate	Operational and process controls
Voice Mail	Anyone	As required to suit message	All employees as appropriate	Message specific
CC:Mail; E-Mail	Anyone	As required to suit message	All employees as appropriate	Message specific
Training Courses	Line Managers/Supervisors	Specific job or task activities	All employees as appropriate	Proper job functioning; consequences of deviations; parameters
Design Reviews	Project Manager	Process/product development	Mgmt; technical staff; QA; EHS; etc.	Technical; market; finance; QA; EHS; etc. reviews

*   Continuous Improvement Team
** Corrective Action Team

methods shown, a manager can use them to discuss both quality and environmental issues, but does not necessarily need to do so. It is obvious that a large number of meetings will focus on either one or the other because due to some critical issue that has surfaced (e.g., continued overdues or a regulatory agency inspection).

Let's now turn our attention outside the organization …

11.4 EXTERNAL COMMUNICATIONS

Communication with the public, other organizations outside the company, and regulatory agencies has been gaining in importance as the public continues to express its concern over industry's pollution control activities and the awareness of a business' impact on the local community in the event of a business or natural disaster. Company stockholders have also recognized that a business' impact on the environment and the surrounding community can strongly influence a company's overall profit level and, thus, stock values. Additionally, the type of information that the media can obtain may damage a company's reputation and, thus, this information must be effectively managed. External agencies must be handled in a proactive manner to hopefully minimize any potential future regulatory fines and citations.

External communication covers a broad range and includes the following stakeholders:

- Vendors and suppliers
- Customers
- The neighboring community
- Regulatory agencies
- Nongovernment organizations (NGOs)
- The media
- Stockholders

You will find that on a day-to-day basis, most of your external communication will be with vendors, suppliers, and customers and then with the regulatory agencies and NGOs, and then the neighboring community, the media, and the stockholders. We will be addressing all of these stakeholders during this section.

If you review Table 11.1 at this time, you will notice that the primary "overlap" between the two standards is in the area dealing with suppliers and contractors and customers. For most companies it is in the area of quality that most of the advance work in dealing with these three entities has taken place. Customer Service organizations generally deal with three types of communication from vendors, suppliers, and customers: inquiries, complaints, and questionnaires. I highly recommend that a procedure be developed to handle complaints separately from inquiries and questionnaires. The intent is to make the Customer Service organization the primary "clearinghouse" or focal point for all inquiries, questionnaires, and complaints — no matter what the content is. A good procedure will detail the desired flow of information and response and the responsibilities of all concerned parties within the organization. It is with this in mind that we will focus our efforts for the integration process.

11.4.1 INQUIRIES AND QUESTIONNAIRES

Many requests for information typically focus on a company's organizational profile, its policy, mission statement, core values, etc., but may also be more explicit in terms of what a company's objectives and targets may be.

A procedure to handle both quality and environmental inquiries and questionnaires is very easy to develop and you will find (if not already) that many inquiries and questionnaires contain a blend of quality and environmental questions. Thus, it makes sense to develop one procedure to handle the process.

The scope of such a procedure can cover miscellaneous surveys, inquiries, and questionnaires, environmentally related questionnaires, and other external communiqués. A typical list of topics being addressed may include some of the following:

- NAFTA or other trade-related issues
- Quality and EH&S policies (or a combined operational policy)
- Any significant impacts
- Use of ozone depleting substances
- Use of hazardous or banned/restricted materials
- The recycling of packaging and shipping containers to customers
- General reuse of materials
- Final product recyclability
- Certifications (i.e., ISO 9001, ISO 14001, etc.)
- Product hazard information
- Supplier assessment audits
- Business recovery plans
- Capacity verification
- Quality plans
- Waste minimization plans
- Country of origin affidavits
- Environmental Performance Evaluations (EPEs)

After identifying the types of inquiries or questionnaires, what needs to be done is to establish a procedure which defines responsibilities and how the inquiry will "flow" through the process. Additionally, a response process must be included. Rather than spending time writing a detailed description of the process, Figure 2 provides a flow chart that will give you an idea on how you may develop your own procedure.

Now what is needed is some process to track and ensure the inquiries are being addressed in a reasonable timeframe. What you will observe in Figure 3 is a "Response Request Form" and in Figure 4 is the database that assists the Customer Service group to monitor the progress of the inquiry. Both of these forms will help Customer Service keep a "handle" on things.

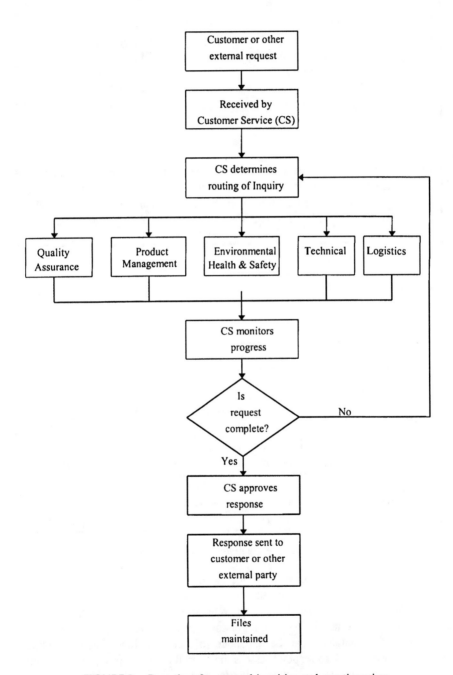

FIGURE 2. Procedure for external inquiries and questionnaires.

Date of Request: (date)   Return Request to: _____
Return by: (date)   Response No.: _____
Source of   ❑ Customer   Type of   ❑ Quality Assurance
Inquiry: ❑ Supplier   Inquiry: ❑ Logistics
             ❑ Regulatory Agency   ❑ EH&S
             ❑ Community   ❑ Product Mgmt
             ❑ Media   ❑ Technical
             ❑ Stockholder   ❑ Other
             ❑ NGO
             ❑ Other

Attached is an inquiry/questionnaire which the Customer Service group has
determined requires the attention of your department. Please return this form
and your response by the date indicated above. If you are unable to respond
in the specified time, please contact the Customer Service representative indicated above.

FIGURE 3.   Inquiry response request form.

No.	Requester	Date Received	Routed to:	Due Date	Response Received
98-001	Acme Chemical	3/31/97	Marketing	4/10/97	4/5/97
98-002	ABC Co.	4/15/97	QA	4/25/97	4/27/97
98-003	XYZ Electronics	4/17/97	Technical	4/27/97	4/20/97
98-004	U.S. EPA	6/22/97	Env. Dept.	7/2/97	7/1/97
98-005	A&C Distributor	7/05/97	Logistics	7/15/97	7/27/97
98-006	SEMARNAP	7/30/97	Env. Dept.	8/10/97	8/5/97
98-007	KXYZ TV	8/08/97	Public Relations	8/18/97	8/15/97
98-008	Home Association	9/10/97	Public Relations	9/20/97	9/22/97

FIGURE 4.   Sample inquiry/questionnaire tracking log.

As you can see, this form need not be too long, but will assist in helping
respective groups to keep the necessary attention on a request for information. This
is particularly important when dealing with requesters who can potentially "damage"
a company (e.g., the media, NGO's, regulatory agencies, etc.).

This kind of database can be expanded to include columns which total the
number of days it took to respond, assess the overall response time, and determine
how each responding department is doing in meeting the time requirements. All of
these metrics, of course, should be used to drive and improve continual improvement.

One other point that must be included in your procedure concerns what to do if an inquiry is received by someone other than the Customer Service group. Specifically, it is important for all personnel to understand this procedure and to be aware of the requirement to route any inquiry or other request for information directly to the service organization which has the responsibility for managing the process. This is especially critical when dealing with complaints.

11.4.2   CUSTOMER AND OTHER EXTERNAL COMPLAINTS

Compared to complaints, the procedure for handling inquiries and questionnaires is relatively simple. The reason for this is that complaints will generally result in some form of corrective action that may or may not involve company liability. Below is a typical list of topics found on inquiries and questionnaires:

- Toxic releases from products or other product safety concerns
- Community noise or concern over environmental releases
- Use of hazardous or banned/restricted materials
- Disposal problems with product or packaging
- Product reliability
- Product performance
- Product functionality
- Late shipment

There is no reason why the process for handling these types of complaints should be much different from that established for basic inquiries and other requests for information. The main difference is in how complaints are managed through the process. Specifically, this may mean much more restrictive requirements for response time, the application of various problem solving techniques, and the development of a review committee to assess the particular issue.

Many companies, in fact, have a standing review board which addresses both internal and externally-related corrective actions. It may additionally require the Customer Service group to establish a customer response procedure as a means to assure complainants are receiving immediate attention. Figure 5 shows a basic process flow for a complaint and, as you can see, it is a bit more complicated than that for an inquiry.

As mentioned in the previous paragraph, the Customer Service group should have a procedure in place that also includes the acknowledgment of the complaint. To that end a form letter should be developed to handle this requirement. Figure 6 contains a sample of a form letter template that can be used to address all issues received.

Complaints, of course, require inclusion into a corrective action process due to the potential liability issues and the need to maintain credibility. The Response Request Form in Figure 3 can be modified slightly and used for handling the complaints as shown in Figure 7.

It is obvious that with a little bit of manipulation and ingenuity, someone can combine the forms in Figure 3 and Figure 7 to create one form that the Customer

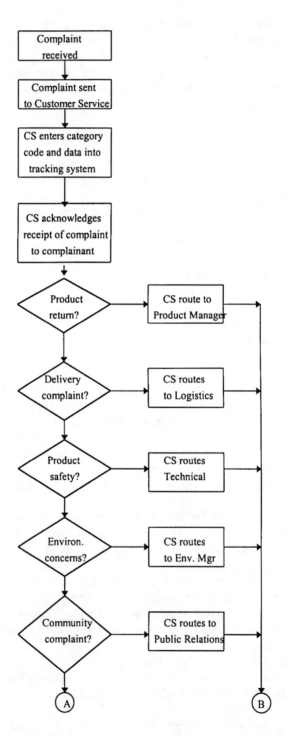

FIGURE 5. Process for customer and other external complaints.

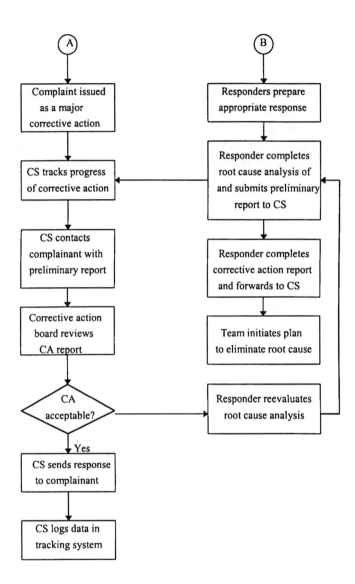

FIGURE 5 (continued).

Service group can use for inquiries, questionnaires, and complaints. Additionally, the tracking log in Figure 4 can be used to track complaints.

11.5   WHAT AUDITORS WILL LOOK FOR

An auditor will look for evidence of processes to communicate both internally and externally. The guideline document, ISO 14004, offers some insightful questions into what auditors may look for:

XYZ Corp.
101 Main St.
Somewhere, USA

Dear (*name*):

This letter is being sent to you to acknowledge XYZ Corp.'s receipt of your complaint regarding (*describe*). Your information and request has been turned over to the following group which has begun an investigation into the cause of the problem:

    ❏ Quality Assurance   ❏ Product Management
    ❏ Logistics   ❏ Technical
    ❏ EH&S   ❏ Other (describe)

In order to complete the investigation, however, we have need of the following additional information:

    ❏ Additional samples   ❏ Product failure data
    ❏ Your testing results   ❏ Product performance data
    ❏ Suspected environmental releases   ❏ Packaging concerns
    ❏ Suspected health concerns   ❏ Other (describe)

The final corrective action describing the cause(s) and corrective action implementation date will be sent to you within (#) days from final receipt of all needed information.

We would like to thank you for assisting us in improving our customer service performance and the quality of our product and services. Feel free to contact me with any additional comments or suggestions.

Sincerely,

FIGURE 6.   Complaint acknowledgment letter.

- What is the process for receiving and responding to the organization's employee concerns?
- What is the process for receiving and considering the concerns of other interested parties?
- What is the process for communicating the organization's quality and environmental policies and performance both internally and externally?
- How are the results from QMS and EMS audits and reviews communicated to all appropriate people in the organization?
- Are internal communications adequate to support continual improvement?

Date of Complaint: (date)	Return Response to:_____

Date of Complaint: (date)   Return Response to:_____
Return by: (date)   Response No.: _____
Source of   ❏ Customer   Type of   ❏ Quality Assurance
Complaint: ❏ Supplier   Complaint: ❏ Logistics
               ❏ Regulatory Agency   ❏ EH&S
               ❏ Community   ❏ Product Mgmt
               ❏ Media   ❏ Technical
               ❏ Stockholder   ❏ Other
               ❏ NGO
               ❏ Other

Attached is a complaint that the Customer Service group has determined requires the attention of your department. Please return this form and your response by the date indicated above. If you are unable to respond in the
specified time, please contact the Customer Service representative indicated
above.

FIGURE 7.   Complaint response request form.

An auditor will ask for evidence which supports the questions listed above.
Copies of customer inquiries, questionnaires, and complaints should be on file and
there should be responses and investigative reports accompanying them. In some
cases, such documents are generally maintained by an appropriate product manager
which has the tendency to "spread" the information out and make it difficult for an
auditor to find what he wants in a timely manner. As an environmental manager, I
have found it expedient to copy all reports I have been involved with and maintain
a separate file strictly for the purposes of ISO 14001.

12 EMS Documentation

12.1 INTRODUCTION

This section of the ISO 14001 Environmental Management Standard, Element 4.4.4, Environmental Management System Documentation, concerns the requirement to *"establish and maintain information, in paper or electronic form, to (a) describe the core elements of the management system and their interaction;"* and *(b) "provide direction to related documentation."*

The comparative requirement under the ISO 9001 Standard can be found in Section 4.2.1, *General*:

> The supplier shall establish, document and maintain a quality system as a means of ensuring that product conforms to specified requirements. The supplier shall prepare a quality manual covering the requirements of this International Standard. The quality manual shall include or make reference to the quality system procedures and outline the structure of the documentation used in the quality system.

As you can see in the ISO 9001 Standards, there is a very explicit requirement to prepare a quality manual covering the requirements of the ISO 9001 Standards. It seems logical to use this format to meet the requirements of the ISO 14001 Standards as well. It will be the subject of this section of the chapter to show you how this can be accomplished and how to develop one operational manual that describes both the quality and environmental management systems.

12.2 THE SCOPE OF QMS AND EMS DOCUMENTATION

Before we begin the development of the manual, it is important to first understand just what the scope of the QMS and EMS documentation entails so that we can see how they interact. The best way to do this is to go through the two standards and systematically list what documentation is required. Table 12.1 lists the ISO 14001 documentation and Table 12.2 lists the ISO 9001 information needing documentation. You will note the key words throughout both of the standards indicating the need to *record*, *identify*, and *document* various types of information. As you can see, the amount of information can be rather formidable.

As was discussed in Section 4.3, you can see why some management is reluctant to implement ISO 14001 because of the potential document burden on the organization. Taken and implemented separately, the two ISO standards can and will be a burden. However, as this book is intending to show, the integration of the two standards can very effectively diminish the amount of documents overall, especially in the area of procedures.

TABLE 12.1
ISO 14001 Documentation Requirements

Section	Documentation
4.2	• *Documented* environmental policy
	• *Records* indicating communication of policy to employees
4.3.1	• *Procedure* for identifying aspects
4.3.2	• *Procedure* for identifying and accessing legal and other requirements
4.3.3	• *Documented* objectives and targets
4.3.4	• *Documented* environmental programs
4.4.1	• *Documented* roles, responsibility and authority
4.4.2	• *Procedure* for identifying training needs
	• *Records* indicating communication of impact of work activities and consequences of departing from operating procedures
4.4.3	• *Procedure* for internal communication
	• *Procedure* for external communication
	• *Records* for consideration of external communication of significant aspects
4.4.4	• *Information* on core elements of management system
4.4.5	• *Procedure* for controlling all documents
4.4.6	• *Identification* of operations and activities associated with significant aspects
	• *Procedures* covering situations where absence could lead to deviations from policy, objectives, and targets
	• *Procedures* related to identifiable aspects of goods and services used
4.4.7	• *Procedure* to identify potential and response to emergency situations
4.5.1	• *Procedures* to monitor and measure key characteristics of operation
	• *Recording* of information to track performance, relevant operational controls, and conformance to objectives and targets
4.5.2	• *Procedure* for defining responsibility and authority for handling and investigating nonconformance, taking action to mitigate impacts, and initiating and completing corrective and preventive action
	• *Record* of any changes to documented procedures
4.5.3	• *Procedure* to establish and maintain records
	• *Document* retention times
4.5.4	• *Procedure* for audits
	• *Records* of audits
4.6	• *Records* of management reviews of:
	— Environmental management system
	— Possible changes to policy
	— Possible changes to objectives and other elements of EMS
	— Audit reports

TABLE 12.2
ISO 9001 Documentation Requirements

Section	Documentation
4.1.1	• *Documented* quality policy, objectives and commitment to quality
	• *Records* indicating communication of policy to employees
4.1.2.1	• *Documented* responsibility, authority, and interrelation of personnel
4.1.2.2	• *Identify* resource requirements
4.1.2.3	• *Identify* management representative
4.1.3	• *Records* of management reviews of:
	— Quality management system
	— Possible changes to policy
	— Possible changes to objectives and other elements of QMS
	— Audit reports
4.2.3	• *Document* how the quality requirements shall be met
4.3.1	• *Documented* procedures for contract review
4.3.3	• *Identify* how a contract is amended
4.3.4	• *Records* of contract reviews
4.4.1	• *Documented* procedure to control and verify design of product
4.4.2	• *Documented* plans of design and development activity
4.4.3	• *Identify* organizational and technical interfaces
4.4.4	• *Identify and documented* product requirements
4.4.5	• *Documented* design output
	• *Documented* reviews of design output documents
4.4.6	• *Records* of design reviews
4.4.7	• *Record* of design verification
4.5.1	• *Documented* procedure to control all documents and data
4.5.2	• *Documented* procedure identifying current document revision status
4.5.3	• *Records* of document and data changes
4.6.1	• *Documented* procedure to ensure purchased product conforms to requirements
4.6.2	• *Records* of acceptable subcontractors
4.6.3	• *Records* of purchase documents
4.6.4.1	• *Records* of product verification and release
4.7	• *Documented* procedure for control of customer-supplied product
4.8	• *Documented* procedure for identifying product receipt, delivery, and installation
	• *Documented* procedure to identify individual product or batches
4.9	• *Documented* procedures defining manner of production, installation, servicing
	• *Records* of compliance with standards/codes, etc.
	• *Records* of process parameter monitoring and controls
	• *Identification* of requirements for any process qualification

TABLE 12.2 (continued)
ISO 9001 Documentation Requirements

Section	Documentation
4.10.1	• *Documented* procedure for inspection and testing activities
4.10.2.2	• *Recorded* evidence of time exercised at subcontractor premises
4.10.2.3	• *Records* of incoming product released for urgent production
4.10.5	• *Records* providing evidence product is inspected and/or tested
4.11.1	• *Documented* procedure for control, calibration, and inspection of measuring and test equipment
	• *Records* of inspection for test software or comparative references
4.11.2	• *Identify* all inspection, measuring, and test equipment
	• *Identify* process employed
	• *Identify* inspection, measuring and test equipment calibration status
	• *Records* of calibration
	• *Records* of out of calibration
4.12	• *Identify* inspection and test status of product
4.13.1	• *Documented* procedure for handling nonconforming product
4.13.2	• *Identify* responsibility for review and authority for dispositioning nonconforming product
	• *Record* of customer acceptance of nonconforming product
4.14.1	• *Documented* procedure for implementing corrective and preventive action
	• *Records* of changes to procedures resulting from corrective and preventive action
4.15.1	• *Documented* procedure for handling, storage, packaging, preservation and delivery of product
4.16	• *Documented* procedure for identifying, collecting, indexing, access, filing, storage, maintenance, and disposition of quality records
4.17	• *Documented* procedure for internal quality audits
	• *Records* of the internal quality audits
	• *Records* of follow-up audit activities and implementation of corrective action
4.18	• *Documented* procedures for identifying training needs
	• *Records* of training
4.19	• *Documented* procedure for performing, verifying, and reporting servicing
4.20.1	• *Identify* the need for statistical techniques
4.20.2	• *Documented* procedure to implement and control the application of statistical techniques

12.3 THE OPERATIONAL MANUAL

As ISO 9001 had indicated, there is a requirement for the development of a manual that describes the elements of the QMS per the International Standards (guidance on the manuals is found in ISO 10013). This manual can be used to meet certain requirements of the Environmental Management System as well.

My intent in this section is not to provide an example of an operational manual. I do not intend to go step-by-step through each section of a quality manual and show how the environmental requirements can be integrated into it — this book provides all the examples which you will need when you begin to develop your own manual. For instance, when defining Responsibility and Authority, you need look no further than Section 9.2 for specific examples. You must be aware, however, as you develop the manual that ISO 14001 does contain certain requirements not found in ISO 9001. These distinctions have been previously discussed, but will be listed again:

- The inclusion in the policy and the discussion as to how the system will maintain *"continual improvement and prevention of pollution"*;
- The requirement to identify *"significant impacts"*;
- The requirement to *"establish and maintain (environmental management) programs for achieving its objectives and targets"*; and
- "Emergency preparedness and response."

12.3.1 THE DOCUMENT DIRECTORY

Before closing this section, I would like to provide an example for satisfying one of the requirements found under both of the standards — the need to "provide direction to related documentation" (ISO 14001) and to "outline the structure of the documentation used in the quality system" (ISO 9001). One of the best ways to do this is to develop and lay out what I call a "document directory." The intent of the directory is to be able to show how all of the documents within your operational system interrelate and cross reference with each other. Figure 1 shows an example of an environmental and quality document directory.

The column headings reflect certain program areas that are common to either or both ISO 14001 and ISO 9001. Below each heading are document descriptions typical to quality and environmentally related procedures. You will notice that the directory also includes any applicable regulatory requirements and/or the related International Standard.

12.4 WHAT AUDITORS WILL LOOK FOR

As stated at the beginning of this section, an ISO 14001 auditor will be much more impressed if you are able to produce an EMS Standards Manual similar to the Quality Manual, whether or not you have been able to integrate the two into an Operational Manual. Since both standards require some kind of document "road map," it is advisable to develop some form of document hierarchy or directory showing the interrelationship of the entire management document system. It is highly recommended that you start with your policy at the top and gradually dissect your system down to the basic task procedures used by operators to produce your product, then to the metrics and database.

The primary principles of your policy, whether it be just quality-related or environmentally- related, can provide the column headers for the directory. Be sure the directory includes all four document levels: (1) the Standards Manual(s); (2) top level management programs/procedures; (3) task or job procedures; and (4) data and metrics. The primary thing to remember is that "all roads must lead back to the policy!"

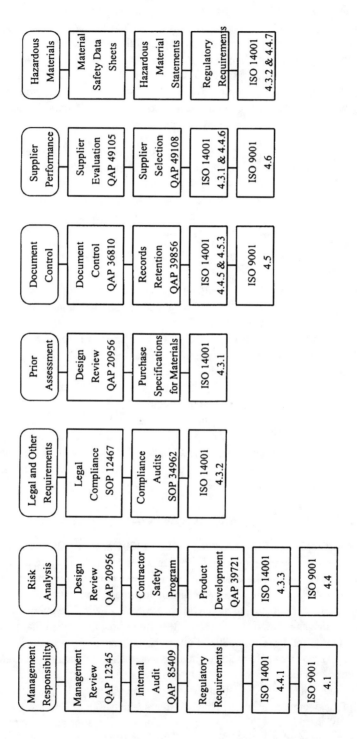

FIGURE 1. ISO 14001 and ISO 9001 directory.

13 Document Control

13.1 INTRODUCTION

This section will concern itself with a topic that is, basically, one of the primary subjects of this book — integrating your ISO 14001 documents into the ISO 9001 document control system. Many of the examples I have shown and will show in the remainder of the book are examples of document integration as well as systems integration. The most likely difficulty you will encounter in your efforts to integrate both of the document systems will be in structuring the overall document system in such a manner that you do not accidentally alter or impede the intent of either of the standards. I think you will find, however, as I have, that the likelihood of this happening is very minimal. The greatest obstacle may be in the reluctance of the quality assurance department in accepting the proposed changes to their system. Their questions regarding the environmental requirements may only be the result of their lack of understanding — this is definitely the time to work together.

Table 13.1 shows the side-by-side comparison of the two standards and, you will note, there are a lot of similarities between the two requirements. The reason for this is that the drafters of ISO 14001 used the quality standards as the template when writing the environmental management system requirements. There will, of course, be some environmental documents that cannot be effectively managed through the quality document control process. We will address this issue in a later section.

13.2 QUALITY CONTROL OF ENVIRONMENTAL DOCUMENTS

A very large percentage of the EMS documents can be controlled by the system established under the ISO 9001 framework. Table 13.2 contains a list of such documents. I think you will notice that this list contains documents from every single section of ISO 14001. The main point of this is that no matter whether you have procedures integrating ISO 9001 and ISO 14001 or the procedure is strictly for ISO 14001, you can have the "quality" document control department manage all of the documents required by ISO 14001. Your quality assurance procedure can be very easily adapted to include the EMS documents. The procedure can be written in such a way as to control the EMS procedures and also refer to the control of other EMS documents.

13.3 DOCUMENT CONTROL PROCEDURE

This section will provide an example of a procedure that addresses the methods to control documents according to the requirements of both ISO 9001 and ISO 14001. The intent is not to provide all of the details of an integrated procedure, but to give

TABLE 13.1

Correlation of "Document Control" Requirements

ISO 9001		ISO 14001	
The supplier shall establish and maintain documented procedures to control all documents and data that relate to the requirements of this Intentional Standard including, to the extent applicable, documents of external origin such as standards and customer drawings.		The organization shall establish and maintain procedures for controlling all documents required by this International Standard to ensure that ...	
Section	**Description**	**Section**	**Description**
		4.4.5(a)	...they can be located.
4.5.2	The documents and data shall be reviewed and approved for adequacy by authorized personnel prior to issue.	4.4.5(b)	... they are periodically reviewed, revised as necessary and approved for adequacy by authorized personnel
4.5.2 4.5.2(a)	A master list or equivalent document control management system are performed status of documents shall be established This control shall ensure that ... the pertinent issues of appropriate documents are available at all locations where operations essential to the effective functioning of the quality system are performed	4.4.5(c)	...the current versions of relevant documents are available at all locations where operations essential to the procedure identifying the current revision effective functioning of the environmental
4.5.2 4.5.2(b)	...and be readily available to preclude the use of invalid and/or obsolete documents. This control shall ensure that ... invalid and/or all points of issue or use, or otherwise assured against unintended use	4.4.5(d)	...obsolete documents are promptly removed from all points of issue and points of use, or otherwise assured obsolete documents are promptly removed from against unintended use
4.5.2(c)	...any obsolete documents retained for legal and/or knowledge preservation purposes are suitably identified.	4.4.5(e)	...any obsolete documents retained for legal and/or knowledge preservation purposes are suitably identified.
		Documentation shall be ...	
4.16	All quality records shall be legible and ...	4.4.5	legible
		4.4.5	*dated (with dates of revision) and*
		4.4.5	*readily identifiable,*
4.16	shall be stored and retained in such a way that they are readily retrievable	4.4.5	maintained in an orderly manner and
4.16	Retention times of quality records shall be established	4.4.5	retained for a specified period
4.16	The supplier shall establish and maintain documented procedures for identification, collection, indexing, access, filing, storage, maintenance, and disposition of quality records	Procedures and responsibilities shall be established and maintained concerning the ...	

TABLE 13.1 (continued)
Correlation of "Document Control" Requirements

ISO 9001		ISO 14001	
		4.4.5	creation and
4.5.3	Changes to documents and data shall be reviewed and approved by the same functions/organizations that performed the original review and approval, unless specifically designated otherwise.	4.4.5	modification of the various types of document

TABLE 13.2
Quality Controlled EMS Documents

ISO 14001 Reference	Description of Document
4.2	Environmental Policy
4.3.1*	Procedure for identifying environmental aspects
4.3.2	Procedure for identifying/accessing legal requirements
4.3.3*	Procedure for establishing and reviewing objectives and targets
4.3.4*	Procedure for achieving objectives and targets (programs)
4.4.1	Defining roles, responsibility, authority
4.4.2	Procedure for identifying training needs
4.4.2	Procedure for employee certification (competence)
4.4.3	Procedure for internal communications
4.4.3	Procedure for external communications
4.4.4	Environmental Management Standards Manual (can be Operational Manual if combined with Quality Manual)
4.4.5	Procedure for Document Control
4.4.6	Standard Operating Procedures
4.4.7	Procedure for Emergency Preparedness and Response
4.5.1	Procedure for Monitoring and Measuring
4.5.2	Procedure for Nonconformance and Corrective and Preventive Action
4.5.3	Procedure for Records Retention
4.5.4	Procedure for Conducting Internal Audits
4.6	Procedure for Management Reviews
* 4.3.1, 4.3.3, and 4.3.4 can be combined into a single "Planning" procedure	

you some highlights that you can use to write a more detailed document control procedure. I have used the list of recommended procedure headings mentioned in Chapter 9 (Section 9.2.7).

Purpose   To ensure all quality and environmental documents within the organization are controlled, implemented, and maintained.

Scope   This procedure defines the methodology for ensuring the control of all documents within the quality and environmental systems of the organization.

Definition of terms

- **Environmental department-controlled document**   A document which is under the specific control of the Environmental Department, Corporate, or regulatory agencies and is not specifically managed by the quality assurance department and does not require signature approval prior to being published and distributed.
- **Controlled document**   A document which is under the specific control of the Quality Assurance department and which requires signature approval prior to being published and distributed. This refers mainly to Quality Assurance Procedures (QAPs) and Standard Operating Procedures (SOPs).
- (Define other terms as needed.)

Referenced documents

- **SOP #12345, Environmental Department Document Control**, which will be discussed in Section 13.4 below.
- (list all documents which may be of importance in relation to this document control procedure, including any regulatory (i.e., Title 29 CFR, etc.) and other standards references (i.e., ISO 14001, Responsible Care, API, etc.)

EH&S precautions   generally "NA" in such a document as this.

Precedence   In the event of a conflict between this procedure and either a written customer requirement, an internal company environmental requirement, a regulatory requirement, ISO 9001 requirements, or ISO 14001, the customer, company, regulatory requirement, or ISO requirement shall take precedence provided it has been approved by the appropriate authority within the company.

Responsibilities

Document Review and Approval

- The document owner is responsible for ensuring his/her document is maintained and kept up-to-date.
- Controlled documents external to the organization, but used by the organization, shall be reviewed and approved by the respective document owner and organization.
  (Example: standards issued by a parent company)

- Document Control shall assign and control all quality and environmental document and revision numbers and historical records.
- Company Environmental documents shall be reviewed and approved by the Environmental Department.
- Document Control shall have the responsibility to distribute all controlled quality and environmental documents."
- "Document Control will maintain the master file of all controlled quality and environmental documents and will provide reference copies on a need-to-know basis.

Document Control Responsibilities and Authority

- All environmentally-related documentation (listed in the procedure section) not directly under the control of the QA document control department shall be managed and controlled by the Environmental Department and identified by a stamp: "**Environmental Department Controlled Document**".

Procedure   This section describes how a new document or a change enters the system including environmentally related.

You will notice that your quality document control procedure needs to be amended very little to include the environmental requirements. This system can very easily support the requirements listed in Table 13.1. What needs to be addressed now are those documents considered within the scope of your EMS, but are difficult to control or cannot be controlled within the scope of the quality document control system.

13.4   OTHER EMS DOCUMENTS

ISO 14001 states in the opening paragraph of Section 4.4.5, that *"the organization shall establish and maintain procedures for controlling all documents ..."* The key thing to note is that "all" documents are included. Other types of documents required and maintained under your EMS may include:

- training records.
- EMS orientation records.
- information on your aspects, objectives and targets, and the accompanying programs.
- manuals of regulations and other standards.
- material safety data sheets.
- manuals of regulatory required programs and accompanying records/data (i.e., hazardous waste manifests, inspection checklists, Hazardous Material Business Plans).
- meeting minutes and records.
- surveys, questionnaires, and inquiries from external sources.
- complaints from customers and other external sources.
- communication memos to suppliers and contractors.

- records of emergency evacuations and drills.
- data logs for monitoring and measuring.
- records of corrective and preventive actions.
- internal and external audit reports.
- records of management reviews.

The question is thus "how do we control these documents?"

13.4.1 ENVIRONMENTAL DEPARTMENT CONTROLLED DOCUMENTS

There are various solutions to this question that can potentially range from very easy to very hard. The solution I am offering is more on the easy end of the spectrum and consists of creating another procedure specific for the environmental documents.

This new procedure identifies what is considered to be an environmentally-controlled document, document distribution, and other document control requirements. We have already mentioned this document in Section 13.3 above as a Reference Document. Let's identify this procedure as SOP 12345, *Environmental Department Controlled Documents,* and, in this case, take the time to write a full procedure:

Purpose   The purpose of this procedure is to define how environmentally related documentation shall be controlled.

Scope   This procedure applies to all documents within the environmental management framework.

Definition of terms

- **Environmental Department Controlled Document**   A document which is under the specific control of the Environmental Department, Corporate, or regulatory agencies and is not specifically managed by the quality assurance department and does not require signature approval prior to being published and distributed. See list above.

Referenced documents

- ISO 14001, Element 4.4.5, Document Control
- ISO 14001, Element 4.5.3, Records
- Title 29 Code of Federal Regulations (list specific requirements covered by this procedure)
- QAP #56789, QA Document Control
- QAP #11111, Records Retention

EH&S precautions   (generally Not Applicable, "NA," for this type of document.)

Precedence   The requirements of this procedure shall only be superseded as a result of changes in regulations, International Standards, and/or internal company requirements.

APPENDIX A
Distribution of Documents

Document Type	EH&S	Corp EHS	Eng.	Mgr	QA	CSC
Training Records	•					
Objectives and Targets	•		•	•		
MSDSs	•	•		•		
Metrics and Data Logs	•		•	•		
HMBP	•	•				
Customer Inquiries, etc.	•				•	•

Responsibilities It is the responsibility of the Environmental Management Department to control, review, and update all documents identified in this procedure according to the requirements set forth by the local, state, and federal regulatory agencies, Corporate Environmental Standards, International Standards, and other codes/standards.

Procedure

- The Environmental Manager shall identify all documents within the scope of this procedure.
- Documents within the scope of this procedure shall be stamped: **Environmental Department controlled document**.
- A distribution list of these documents shall be maintained by the Environmental Department (See Appendix A).
- Records shall be retained according to QAP #11111, Records Retention.

13.5 WHAT AUDITORS WILL LOOK FOR

As I mentioned earlier the key thing to keep in mind is whether or not your system controls *all* documents within the scope of your environmental management system. All too often companies seeking ISO 14001 certification forget about the control of environmental data, metrics, training records, etc. which are not readily managed or controlled by the quality document control system. Although the quality document control department with its accompanying procedures and systems can control a large percentage of the documents within the scope of the environmental management system, it may be necessary (as I have shown) to develop another procedure for managing all of the other documents. The new procedure must also consider issues of distribution and retention.

14 Operational Control

14.1 INTRODUCTION

The area of operational control concerns itself with the management of the day-to-day impact of your operations on the environment. The overall intent is to identify those aspects of your operation that will create a significant impact on the environment and implement controls to manage them. You can see from this opening statement that Operational Control (Sub-element 4.4.6) is influenced by ISO 14001 Element 4.3, *Planning*, and Sub-element 4.4.4, *Training, Awareness and Competence*. This will become more apparent as the chapter progresses.

14.2 ISO 14001 REQUIREMENTS

Element 4.4.6 of ISO 14001 states:

> The organization shall identify those operations and activities that are associated with the identified significant environmental aspects in line with its policy, objectives, and targets. The organization shall plan these activities, including maintenance, in order to ensure that they are carried out under specified conditions by:
>
> (a) establishing and maintaining documented procedures to cover situations where their absence could lead to deviations from the environmental policy and the objectives and targets;
> (b) stipulating operating criteria in the procedures;
> (c) establishing and maintaining procedures related to the identifiable significant environmental aspects of goods and services used by the organization and communicating relevant procedures and requirements to suppliers and contractors.

The key factor (highlighted) stipulated in these requirements centers on the need to implement controls based on whether or not they are in line with and minimize or prevent deviations from the environmental policy and the objectives and targets. This section also indicates the need to concern oneself with the influence and impact of three other operational activities: maintenance; goods and services; and suppliers and contractors.

14.3 COMPARISON WITH ISO 9001

ISO 9001 has several sections where the requirements of ISO 14001 can be integrated. Table 14.1 shows these interrelationships:

TABLE 14.1

Correlation of "Operational Control" Requirements

ISO 9001		ISO 14001	
Element	Requirement	Element	Requirement
4.2.2 (a)	Quality system procedures consistent with quality policy	4.4.6 (i)	In-line with environmental policy and objectives and targets
4.9 (g)	Maintenance of equipment	4.4.6 (i)	Maintenance
4.2.2 (b)	Documented procedures	4.4.6 (a)	Documented procedures (where absence could lead to deviations)
4.9 (a)	Documented procedures (where absence could lead to deviations)		
4.9 (I)	Process control	4.4.6 (b)	Operating criteria
4.9 (d)	Monitor and control process parameters		
4.6.1	Purchasing procedure	4.4.6 (c)	Goods and services
4.6.3	Purchasing data		
4.7	Control of customer-supplied product procedure		
4.3.1	Contract review procedure	4.4.6 (c)	Suppliers and contractors
4.6.2	Evaluation of subcontractors		

(i) indicates first paragraph of element, etc.

(a) indicates subsection "a" of element, etc.

As you can see, the quality system provides several integrating opportunities for ISO 14001. Let's now take each of the ISO 14001 requirements and see how we can specifically integrate them into the quality system.

14.4   MAINTENANCE ACTIVITIES

When discussing maintenance activities, it is important to distinguish between maintenance classified as "routine" vs. nonroutine. "Nonroutine" is considered to be equipment breakdowns that are "unplanned." Routine maintenance is commonly called "preventive maintenance."

Preventive maintenance is "predetermined" maintenance work performed to a schedule without specific preknowledge of any defect. In the Quality Management System it can be an effective tool to combat the development of equipment weaknesses into product defects and unplanned downtime. Preventive maintenance can have both a positive and a negative effect on the system. From a positive standpoint it can be utilized as part of a hazard assessment program for identifying potential hazardous situations and other impacts, including environmental. On the negative side, failure to perform preventive maintenance as required can result in an undesirable impact. This is a clear example of what is meant in the standards where *"absence of such procedures could adversely affect quality (environment)."*

A good example of a negative impact resulting from the absence or failure to implement a required procedure can be readily demonstrated with a piece of equipment which utilizes various chemicals in the manufacture of a product. If we assume the equipment has a pressurized piping system for the chemicals, it becomes obvious that regular inspections of valves and seals are required to prevent not only minor leakage, but also a major rupture in the piping system. If the equipment system is large, a valve or seal rupture could create a major release to the surrounding area and, if a sewer or stormwater system is also nearby, then you also have unauthorized contaminated water discharge.

We can also use the same scenario to demonstrate a "positive" impact. A good, well-managed preventive maintenance system could greatly minimize or even eliminate such a rupture of the system. The best way to ensure that preventive maintenance occurs is to have not only a written procedure (Maintenance Operating Procedure or MOP) of what to do, but also a scheduling system. There are a large number of preventive maintenance software packages available to manage the program. The preventive maintenance procedure can be written in such a manner as to address not only quality issues, but also environmental. An example of a MOP may include some of the following tasks:

- While checking the valves and seals, look to see if there is any evidence of chemical on the floor.
- Ensure temperature gages have been calibrated per the requirements of the Calibration Program.
- When draining vacuum pump oil, be sure to have a spill control kit available and dispose of used oil in the proper hazardous waste container.
- Check all piping surfaces for leaks and corrosion.
- Verify accuracy of pH meter to ensure chemical adheres to proper quality specifications.

The Maintenance Operating Procedure should also include information concerning the "consequences" of not following the procedures. An example of a "consequence" statement may be:

Due to the corrosiveness of the chemical in the piping system, it is critical that the valve seals be replaced on a monthly basis. Failure to replace valve seals on a monthly basis could result in a valve rupture and an unauthorized release of the chemical to the surrounding environment.

One final comment in regards to the inclusion of the maintenance as actual or potential environmental aspect. If you take the time to assess both the negative and positive impacts of maintenance on the quality of your product and on the environment, you may see the need to establish maintenance objectives and targets and subsequent programs to minimize its impact. Table 14.2 shows how all this fits together for our chemical piping system.

TABLE 14.2
Chemical Piping System

ISO 14001 Element 4.3, Planning	Evaluation		
4.3.1 — Aspect	Maintenance of chemical piping system		
4.3.1 — Significant Aspect	Severe release of corrosive chemical		
4.3.2 — Legal and Other Requirements	Potential violation of unauthorized release to sewer system of corrosive chemical		
4.3.3 — Objective	Prevent valve seal ruptures		
4.3.3 — Target	Improve preventive maintenance of system		
4.4.3 — Program/project	What	Who	When
	Investigate PM procedure: Review/modify MOP Review/modify PM schedule	Maintenance	12/31/97 12/31/97
	Evaluate current valve seal material and investigate alternative materials with more corrosion resistance	Process Engineer	12/15/97
	Install automatic shutoff valves with pressure drop sensors every 50 feet to minimize volume of material due to accidental release	Maintenance	1/31/98

14.5 DOCUMENTED PROCEDURES AND OPERATING CRITERIA

ISO 9001, Element 4.2.2, requires that an organization have:

> "the range and detail of the procedures ... dependent upon the complexity of the work, the methods used, and the skills and training needed by personnel in carrying out the activity" with a special note that "documented procedures may make reference to work instructions that define how an activity is performed.

Element 4.9 (a) also adds that

> controlled conditions shall include ... documented procedures defining the manner of production, installation and servicing, where the absence of such procedures could adversely affect quality.

A procedure established under ISO 9001 can thus be an excellent vehicle for meeting the requirements of ISO 14001. It is a very easy task to take an operating procedure associated with an activity or service of your organization and assess whether or not it is an aspect of your operation that may have an actual or potential impact on the environment. Once that determination has been made, it is a simple matter to incorporate information into the procedure which will satisfy the requirements of ISO 14001. Such information may include "awareness and consequence" statements (see Chapter 10), monitoring and measurement requirements (see

Chapter 16), legal and other requirements (see Chapter 7), and employee responsibilities (see Chapter 9).

In order to demonstrate this, I will show you an example of an integrated operating procedure that incorporates both ISO 9001 and ISO 14001 requirements:

Purpose   To ensure that the manufacture of (product) produces high yields with minimal impact on the environment, people, and equipment.

Scope   This procedure defines the control system and specifications for the manufacture of (product).

Definition of Terms

- ARB: Air Resources Board
- VOC: Volatile Organic Compound
- SPC: Statistical Process Control

Referenced documents

- ISO 14001, Element 4.4.6, Operational Control

EH&S precautions

- Safety glasses are required when doing setup
- Powder generated from the waste material shall not be washed or swept into the sewer/drain system
- Hazardous waste material is to be disposed of in accordance with RCRA guidelines
- This process has a permit-to-operate from the ARB. Permit conditions require that a usage log be maintained for the chemical with a total 12-month consumption not to exceed "x" gallons. Failure to maintain the log can result in a violation and/or fine.

Precedence

- The health and safety of personnel and protection of the environment shall take precedence over this operating procedure.
- Customer requests and requirements shall take precedence over this procedure.

Responsibilities

Process Engineer

- For maintaining and revising this procedure as needed.
- For the technical aspects of the operation.

Department management

- For ensuring that personnel operating this equipment are fully trained and certified and that they understand the consequences of not following the procedure.
- For providing all necessary resources to achieve a high quality product and to minimize the EH&S impacts.

TABLE 14.3
Operating Criteria and Controls Comparison

Criteria/Control	ISO 9001	ISO 14001
Product conforms to specified customer requirements	•	
Minimizes the impact on the environment		•
Minimizes the impact on personnel	•	
Reduces the potential for an unscheduled breakdown	•	
Ensures continuing process capability	•	•
Provides sufficient operating time such that preventive maintenance can be scheduled with minimal conflict with overall running time	•	
Monitoring and measurement requirements	•	•
Consequences of deviating from the established procedures	•	•
Emergency shutdown procedures		•
Operator competence	•	•
Other resources	•	•

Environmental Manager
- For maintaining the monitoring data and usage log in compliance with the ARB permit conditions.
- For filing all permit documents and acting as liaison with the ARB.
- For training personnel on all of the environmental requirements.

Maintenance
- For ensuring that all temperature controls are calibrated according to ISO 9001 requirements.

Procedure   The procedure can include both quality and environmental operational controls such as measurements, statistical process controls, etc.

As you can see, this procedure stipulates operating criteria and controls for both ISO 9001 and ISO 14001. Table 14.3 shows which Standard is satisfied by the procedural criteria and controls. As stated on several previous occasions, the operating procedure can be the most valuable tool you have for integrating many of the ISO 14001 requirements.

14.6   GOODS AND SERVICES

We now come to a subject that deals primarily with the control of your purchased materials and supplies. Table 14.1 indicates that the comparative requirements for ISO 14001 under the ISO 9001 standards lie in Section 4.6, *Purchasing*, and in Section 4.7, *Control of Customer-Supplied Product*. Let's take each of these sections and interject where the ISO 14001 requirements might fit in:

Section 4.6.1   The supplier shall establish and maintain documented procedures to ensure that purchased product conforms to specified requirements.

Section 4.6.3 Purchasing documents shall contain data clearly describing the product ordered, including where applicable: (a) the type, class, grade or other precise identification; the title or other positive identification, and applicable issues of specifications, drawings, process requirements, inspection instructions and other relevant technical data, including requirements for approval or qualification of product, procedures, process equipment and personnel, etc.

Section 4.7 The supplier shall establish and maintain documented procedures for the control of verification, storage and maintenance of customer-supplied product provided for incorporation into the supplies or for related activities ...

What has been indicated above and stands out in the ISO 9001 Standards are requirements for product specifications, identification of applicable issues for the process, and the approval of product. Although we spent a great deal of time discussing design reviews in Chapter 6, it is important to address the importance of a design review on materials once again.

14.6.1 ENVIRONMENTAL REVIEWS

In Chapter 6, I discussed the importance of addressing environmental, health and safety issues early on in a design review program in order to proactively eliminate any chemicals or materials which may impact the environment and people. All too often, when a project manager is considering a new or a revised product and/or process, he fails to consider the impact beyond the financial gains for the company. It is critical that an environmental, health, and safety professional be included in the reviews and as early in the process as possible. Too few environmental, health, and safety professionals consider the significant impact of a new chemical or materials beyond the need for a new air permit, for instance. They raise the "red flag" and indicate the product will generate a new air permit, but don't consider the possibility of replacing the material with a non-VOC material and, thus, lessening or minimizing its impact on the environment. An environmental manager who is ISO 14001 "smart" would be an extremely valuable addition to any review team.

Once again, a design review team must address the following type of questions early in the review:

- Will this new or modified process or product create a new environmental aspect?
- Will this new environmental aspect create a significant impact?
- Will this new or modified process or product potentially change or influence a current environmental aspect?
- Will this new or modified process or product potentially change or influence a current significant impact?
- Are there any new regulations that must be taken into consideration with this new or modified product and/or process?
- Is this new chemical/material currently being considered by any of the regulatory agencies for potential reporting requirements, controls, elimination, or restriction in some capacity?

- Is this new chemical/material currently being considered or has been considered for either restricted use or banning within the confines of another country for which you have sales considerations?
- Is this new chemical/material being imported from another country and, if so, are there any potential regulatory issues barring or restricting its use within the confines of your own country?
- Will this new chemical, after its use within your process, pose any restrictions or special considerations for waste disposal, including financial, technological, etc.'?
- Will the purchase of this new chemical or material impact the process by requiring special technological options (e.g., abatement controls) that were not considered during initial design reviews.

Most of these questions provide sound reasoning for evaluating the environmental impacts as early in the design review as possible. Failure to consider them could result in severe financial consequences in terms of legal violations and fines, expensive add-on equipment, and other additional process expenses not considered at all. Additionally, as environmental concern increases, more and more of your customers and other externally interested parties are putting more and more constraints on their suppliers in using chemicals, materials, and substances that impact the environment. Pressure from your customers could result in the need to modify or re-engineer your product using chemicals or materials more friendly to the environment.

14.6.2 PRODUCT SPECIFICATIONS

In conjunction with the design or process review, it is important to purchase supplies and materials that conform not only to your requirements, but also to your customer who may or may not be the final end-user. The best way to do this is through the development of product specification documents that describe the manufacture and acquisition requirements from your own suppliers.

The considerations being documented in the procedure must also go beyond the actual description of the product. Consideration must also extend to the supplier's own handling or manufacturing process in making the product. With ISO 9001 now being so soundly entrenched, it is much easier these days to put pressure on a prospective supplier to manufacture your product in the manner which you delegate or specify. Although I have not been including the ISO 14020 Labeling Standards in this book, the inclusion of them in your product specifications is a critical factor. If the issue of "goods and services" in ISO 14001 is very important to you, it would be highly recommend you become familiar with the series of standards dealing with labeling and packaging (see a list these standards in Table 3.2).

As I have done on several previous occasions, I will provide a sample procedure that acts as a guideline for your own. This procedure will contain, of course, requirements for both ISO 9001 and ISO 14001.

Purpose This purchase specification for material describes the requirements for the manufacture and acquisition of (name of material) from XYZ Co.

Scope   This procedure applies to (name of material) used in the manufacture of widgets.

Definition of terms

- A Certification is a document required from a supplier which guarantees that the material shipped meets the requirements and specifications as stated in this procedure. In cases where specified, it also requires the inclusion of supporting test data or other documents supporting the specifications.
- ARB is the Air Resources Board
- ODS is an Ozone-Depleting Substance
- (Add others as needed.)

Referenced documents

- ISO 14001, Element 4.4.6, Operational Control
- ISO 9001, Element 4.7, Control of Customer-Supplied Product
- ISO 14023, Self Declaration Environmental Claims–Testing and Verification
- Drawing #A17965, Revision C, Widget

EH&S precautions

- The supplier is required to supply Material Safety Data Sheets for each shipment of this material.
- The supplier must have all permits required by the Air Resources Board before manufacturing this material.
- This material must not be manufactured with any ozone-depleting substances.
- All waste generated by the supplier must be disposed or recycled in a manner consistent with the appropriate regulations.

Precedence   In case of a conflict, the requirements in the Purchase Order take precedence over this specification document.

Responsibilities

Process Engineer

- The Process Engineer is responsible for maintaining the technical content of the specification document.

The Supplier

- The supplier is responsible for conforming to all aspects of this specification document.

Environmental Manager

- For verifying that the supplier has the necessary environmental permits and is handling waste in a manner consistent with the required regulations.
- For evaluating the Material Safety Data Sheets and communicating any hazard information to appropriate personnel.

Material Control

- For the proper storage of this material.
- For ensuring shelf life considerations with this material are addressed.
- For ensuring proper labeling and packaging of the material is addressed.
- For inspecting the incoming material to ensure it meets the specifications and contacting the process engineer if there are any discrepancies.

Procedure

Visual criteria

- The color of the material must be (describe).

Material properties

- Describe any physical and chemical properties, such as boiling point, crystallinity, glass transition temperature, viscosity, melting point, pH, etc.

Certifications

- The supplier is required to provide certifications with supporting test data which shows that the material shipped meets the specifications of the properties listed above and will maintain these properties within the shelf life period.

Shelf life

- The material shall meet the specifications for a minimum period of (define period) from the date of shipment when stored at (define storage temperature).

Labeling and packaging

- The supplier shall ensure that the material is packaged and labeled in accordance with all relevant federal, state, and local regulations, including the Montreal Protocol.

14.6.3 Incoming Inspection

The procedure mentioned that Material Control has the responsibility for ensuring incoming material meets the specifications required. What is needed is a procedure to manage the incoming inspection process itself. This is extremely important — failure in this area could cause unnecessary waste and low yields (e.g., an impact on the environment). The requirement for this activity is found in ISO 9001, Section 4.10.2, *Receiving Inspection and Testing*. Rather than writing another procedure at this point, I would like to merely point out some significant information you can use when writing your own.

It is important to note that material being received for incoming inspection should at a minimum include some of the following documents:

- Process FMEA (Failure Modes and Effects Analysis)
- Design FMEA
- MSDS
- Packaging requirements

- Labeling and product identification
- Use of ozone depleting substances (including any "Pass Through" require-
  ments)
- Material characteristics (e.g., physical and chemical)
- Other hazard warning requirements
- Material testing and data reports

If the results of incoming inspection prove to be unacceptable, the inspector or
engineer needs to evaluate why the supplier was unable to meet the material specification. If the supplier cannot meet the specification and the specification cannot
be changed for your process, then a reevaluation of the supplier should be made
with the potential need to find another supplier.

14.7   CONTRACT SUPPLIERS

Now that we have discussed the specifications for purchasing materials, we will turn
our attention to the supplier of the material. Taking the time to find the right supplier
is as critical, if not more so, than finding the right material to do the job. A poor
supplier can cause potential delays in delivery of the material, ship a poor quality
material or provide a material that is inconsistent in its quality. All of these will
adversely affect your own operation through unexpected slowdowns or shutdowns
of your manufacturing process, poor product yields, and a high rate of waste.

What is needed now is a procedure to define the requirements for suppliers:

Purpose   This procedure defines the minimum quality and environmental requirements for suppliers providing materials and/or services.

Scope   This procedure applies to all actual or potential suppliers of material.

Definition of terms

- A *Certification* is a document required from a supplier which guarantees
  that the material shipped meets the requirements and specifications as stated
  in this procedure. In cases where specified, it also requires the inclusion of
  supporting test data or other documents supporting the specifications.
- An *Approved Supplier List* is a list of all suppliers that can be used in
  the purchase of materials.
- A *Critical Commodity* is a material in which a supply disruption would
  have a large negative impact on sales (i.e., sole source, impact of supply
  disruption, difficulty of qualification or requirement of customer) or a
  significant impact on the environment.
- An *ODS* is an Ozone-Depleting Substance.
- (Add others as needed.)

Referenced documents

- Procedure for Material Specifications (See Section 14.6.2 above.)
- ISO 9001, Section 4.6.2, Evaluation of Subcontractors
- ISO 14001, Section 4.4.6(c), Operational Control

EH&S precautions

- The supplier is required to provide Material Safety Data Sheets for each shipment of material if not classified as an "article."
- The supplier must have all operating permits required by appropriate regulatory agencies before material can be produced.
- The material must not be manufactured with any ozone-depleting substances.
- All waste generated by the supplier must be disposed or recycled by the supplier and in a manner consistent with the appropriate regulations.

Precedence   In case of a conflict, the requirements in the Purchase Order take precedence over this qualification document.

Responsibilities

Supplier

- The supplier is responsible for meeting or exceeding all applicable requirements in this document as well as any associated material specification documents for the supplied material. The quality assurance process for the supplier shall be documented and shall be subject to review, evaluation, and audit by the customer.

Environmental Manager

- For verifying that the supplier has the necessary environmental permits and is handling waste in a manner consistent with the required regulations.
- For evaluating the Material Safety Data Sheets and communicating any hazard information to appropriate personnel.

Purchasing is responsible for controlling the supplier evaluation and selection program and maintaining all documents concerning the ongoing evaluation of the suppliers.

Procedure

To add to the approved list

- Requirements for new suppliers include a positive financial profile, a suitable and effective quality management process, compliant Environmental, Health and Safety Systems (see Appendix H for an example) and process capability suitable for the material supplied.
- Supplier evaluation shall follow the ISO 9001 and ISO 14001 series of standards.

To remain on the approved list

- All suppliers are required to maintain satisfactory performance in conformance with quality, environmental, and delivery requirements.

To be removed from the approved list

- A supplier will be removed from the list based on a review of the quality, environmental, and delivery performance.

This procedure does not take into full account all of the minor details required under the ISO 9001 Standards due to the extensive complexity of the requirements. The intent above was to show how some of the ISO 14001 Operational Control requirements can be integrated into various procedures of ISO 9001.

14.8 ONSITE CONTRACTORS

The focus in this section will be on those "contractors" who perform a service directly at your facility. The communication required under ISO 14001 should be in two directions: (a) the contractor informing you of the work he performs which may affect an aspect of your operation and impact the environment (actual or potential); and (b) your need to inform the contractor of any aspects of your operation which may impact him.

When you sign a contract with another business to perform a service directly at your site, it is important to establish certain criteria within the contract itself. This criteria may include information on their EH&S history, the nature of any chemicals which may be used by them (with the provision of MSDSs to you), what waste may be generated from their activities and which aspect(s) of your operation may be affected. Any work that they perform may adversely affect your operation and create a significant impact within your own premises. You should also not forget any potential impacts on the surrounding community. Steps must therefore be taken beforehand to understand where this impact will be and to minimize or eliminate them, if possible.

When evaluating the impact, you also need to consider the manner in which the impact may affect you. The impact may not be obvious, such as the generation of hazardous waste, but may create a large or small scale change on one of your objectives with its resulting targets and programs. It may also impact your ability to comply with certain regulatory requirements or other standards for which there must be adherence.

One of the best ways to manage contractors on your premises is to develop a Contractor Safety Program. The primary focus of such a program is to "inform and train" prospective contractors on the aspects of your operation and to establish some guidelines for communication. The program contains provisions for the distribution of various "special work permits" that the contractor must obtain before beginning work. To obtain the permit, the contract workers must read about specific company environmental and safety provisions or rules pertinent to the particular work to be done. They must then sign a statement of understanding and describe the details of the work to be done. This management system can protect the environment, people, and property quite effectively. It is very important to communicate to prospective contractors the requirement to participate in your Contractor Safety Program as a condition of signing and accepting a final contract.

14.9   WHAT AUDITORS WILL LOOK FOR

An ISO 14001 auditor will look primarily for written operating procedures which should show whether or not the organization has *control* of its processes. They will especially be concerned about the presence of operating procedures that are directly related to any of your identified significant aspects which may very well include maintenance activities. In addition, it is important to demonstrate that personnel understand the consequences of not following the procedures. It is highly recommended that such statements be included in the operating procedure. Operator certification examinations can then include questions pertaining to the consequences of deviating from the established procedure.

The second major concern for the auditor will be how well you handle suppliers of materials, joint venture operations, and onsite contractors. As already mentioned, they can have an impact on your operation and the environment.

15   Emergency Preparedness and Response

15.1   INTRODUCTION

This particular requirement under ISO 14001 is one of the major areas that does not have any direct correlation with ISO 9001. The inclusion of this in the standards, however, is very important due to the fact that the lack of such a program can potentially have disastrous consequences on people, the environment, and property. Because of this and since my primary intent in this book has been to provide practical suggestions for implementing ISO 14001, I will not skip this section of the standard.

The title of this chapter indicates that there are two major requirements in this portion of ISO 14001: (a) to be prepared; and (b) to have a response system in place. But first, let's look at all of the requirements for this section of ISO 14001.

15.2   THE ISO 14001 REQUIREMENT

Element 4.4.7 states:

> The organization shall establish and maintain procedures to identify potential for and respond to accidents and emergency situations, and for preventing and mitigating the environmental impacts that may be associated with them. The organization shall review and revise, where necessary, its emergency preparedness and response procedures, in particular, after the occurrence of accidents or emergency situations. The organization shall also periodically test such procedures where practicable.

The standard has several requirements which need particular attention: first, you need to have a written procedure describing what potential accidents or emergencies you may face; second, the procedure must describe how you plan to prevent them; third, the procedure must describe how to contain or mitigate it if one occurs; fourth, there needs to be a review/revision system; and fifth, the system must be periodically tested. There is also one other requirement found in ISO 14001 not directly addressed in Section 4.4.7, but has direct bearing on the "preparedness" aspect of the requirements. It is found in Section 4.4.2, *Training, Awareness and Competence:*

> It shall establish and maintain procedures to make its employees or members at each relevant function and level aware of ... (c) their roles and responsibilities in achieving conformance with the environmental policy and procedures and with the requirements of the environmental management system, including emergency preparedness and response requirements ..."

It also is important to note that you need to look beyond those aspects that will actually cause an environmental accident and consider situations that have the potential for a chemical spill or air emissions release. For instance, earthquakes, fires, bomb threats, criminal activity and gas leakages are examples of accidents and emergencies that have the potential to impact the environment.

15.3 THE EPRP DOCUMENT

As I have done on several previous occasions, I will provide an example of a procedure that takes into account the ISO 14001 requirements. For the sake of this example, I will reference the various California regulations pertaining to our requirements. However, due to the fact that an Emergency Preparedness and Response Plan document can end up being quite extensive, I will try to limit my information to the primary issues at hand (e.g., meeting the intent of ISO 14001) and allowing you to fill in your company's appropriate information. Additionally, I will not go into all of the details of what personnel should do during certain types of emergencies (such as seeking cover during an earthquake). The main focus of this procedure will be to provide information on how to "prepare" and "respond" to an accident or emergency.

15.3.1 PREPAREDNESS ACTIVITIES

The preparation for accidents or emergencies can be managed through several different and concurrent avenues. These may include:

a. *Orientation classes for new employees:* An orientation most likely will be the initial exposure of an employee to the program. By including basic emergency evacuation procedures, first responder chemical spill requirements, initial hazard communication training and an area-specific tour that includes emergency response equipment, you will have given your new employee a good foundation to be built on.

b. *Hazard evaluations:* Before you can assess what type of emergency systems and equipment you will need, it makes sense to first understand what types of hazards your operation has. Knowing which aspects of your operation will create a significant impact on people, the environment, and property will provide the right information and data to make the assessment (note the connection to Element 4.3 in ISO 14001). The hazard evaluation information should also form the basic information you will communicate in the new hire orientations and the staff and department meetings.

c. *Emergency systems evaluations and surveys*: After you have assessed the type of hazards being faced, you will have a much better idea of what type of emergency response equipment you will need. By utilizing the expertise of outside consultants (i.e., the fire department), you can know exactly where and how much response equipment you will need. It also is important to conduct regular inspections of the equipment to ensure they are in working order and fully available.

d. *Training*: Once you have the equipment in place you can train appropriate personnel (or all of them). Basic and periodic training on such things as fire extinguisher use and chemical spill containment will go a long way in preventing a significant impact.

e. *Staff or department meetings*: Regular discussions of specific accident and emergency issues will keep personnel aware of their responsibilities and roles in the program.

f. *Emergency scenarios and the implementation of practice drills*: These are the best way to see if personnel truly understand their roles and responsibilities and if the system itself functions as it should. Emergency scenarios should not be limited to just pulling the fire alarm, but should also include such things as staging a chemical spill "accident," as well.

Various parts of the information that have just been presented can be included in the sample procedure presented in the next section.

15.3.2   RESPONSE ACTIVITIES

This procedure is formatted a bit differently from the previous procedures presented. Much of the information regarding the types of accidents and emergencies are shown as appendices to the actual procedure.

Purpose   The purpose of this procedure is to define the requirements for preparing personnel to protect the health and safety of themselves and the environment and to appropriately respond in the event of an earthquake, fire, hazardous material or waste release, gas leakage, bomb threat, criminal activity or other medical emergencies (Appendix "X"). One aspect of the overall program is to train personnel on preventive (proactive) techniques as opposed to reactive.

Scope   This procedure applies to all personnel within the organizational site.

Definition of terms

- The *Shift Emergency Coordinator* (SEC) is the lead responder on each operating shift.
- The *Emergency Response Team* (ERT) is a team of personnel who are specifically trained to respond to accidents and emergencies.
- *Material Safety Data Sheets* provide information on the physical and health hazards of chemicals, the required protection systems, and the emergency response and disposal procedures.
- The *Chemical Response Team* (CRT) is trained to respond to a chemical spill of less than (x) gallons.
- The *Hazardous Material Inventory Statement* (HMIS) provides information to the local fire department with information regarding area-specific chemical inventories and hazards during a response.

Reference documents

- Title 8 California Code of Regulations, Section 3220, Emergency Action Plan
- Title 8 California Code of Regulations, Section 3221, Fire Prevention Plans

- ISO 14001, Element 4.4.7, Emergency Preparedness and Response
- Title 40 Code of Federal Regulations, Section 265, Subpart D
- Title 8 California Code of Regulations, Sections 2267140-2267145

EH&S precautions   Specific precautions are dealt with each individual accident or emergency in the appendices.

Precedence   Procedures in this document shall be superseded by internal company requirements, state and federal regulations and the Uniform Fire Code (UFC).

Responsibilities

- The Senior Manager shall appoint an Emergency Coordinator.
- The Emergency Coordinator is responsible for the development, implementation, evaluation, and maintenance of the EPRP.
- Arrange and/or provide appropriate training for ERT and CRT members.
- Ensure ERT meeting schedules are being conducted.
- Ensure shift drills are being conducted as scheduled.
- Update the EPRP under the following circumstances:
  — at least annually
  — after each accident or emergency as needed to ensure corrective actions are implemented to prevent reoccurrence
  — per changes in internal company requirements, the UFC, or state and federal regulations.
- Emergency Response Team members must be trained in the following accident and emergency activities:
  — accident/incident investigations
  — small chemical spill response
  — proper emergency drill activities (see Appendix "X")
  — emergency communications system —(add additional as needed)
- Chemical Response Team must know:
  — the health hazards, physical hazards, and protective equipment required for all chemicals that may require spill clean-up
  — the location, purpose, and application techniques for spill response equipment and materials and maintain adequate response supplies.
  — requesting the assistance of a contracted spill cleanup responder as needed.
- All Other Personnel
  — be familiar with all exits and evacuation routes
  — know their respective muster point location
  — know the location of emergency equipment in their work area and how to use it (e.g., fire extinguishers, emergency eyewash and showers, and small chemical spill kit materials)
  — know emergency shutdown procedures for the equipment they are responsible for.

Tools   N/A

Materials   N/A

Procedure(s)

- Specific accident and emergency procedures are detailed in Appendix "X."
- All probable emergencies are listed in Appendix 'X' which may be experienced by the organization. A Risk Score was assigned to all probable accident and emergencies based on the severity, likelihood and frequency (RS = S $\times$ (L $\times$ F)) where L $\times$ F is the probability of occurrence.

APPENDIX 'X-1' — EARTHQUAKES

After the earthquake shock has subsided and personnel have been told to evacuate, the ERT should check for potential fallen debris, equipment, etc. which may have been knocked down or moved, chemical spills, and any gas odors.

Employees must report all injuries, broken utility lines, fires, chemical spills, gas odors, etc. Do not light matches, smoke, turn on light switches, use any electrical outlets, or electrical appliances!

APPENDIX 'X-2' — ONSITE CHEMICAL RELEASES

This response procedure is to be followed in the event of an unplanned sudden or nonsudden release of a hazardous material or waste to the air, soil, or surface water. By definition, any hazardous material spill becomes a waste. Do not dispose of any spilled material as general trash.

This spill response procedure references items found in the company's Contingency Plan, which has been developed in compliance with the Resource Conservation Recovery Plan (RCRA), Title 40 Code of Federal Regulations, Section 265, Subpart D.

Immediately stop the source of the discharge, *only if it is safe to do so*. This may involve: shutting off equipment or pumps; plugging a hole in operating equipment or a tank; closing a valve; or righting an overturned container or piece of operating equipment.

Contain the release, *only if it is safe to do so*, by using one of the following means: for relatively small spills (<5 gallons), apply absorbent material to the surface of the spill and continue to re-apply until the spill is totally absorbed or contained. For larger spills (>5 gallons), use absorbent material to construct a containment dike around the perimeter in order to prevent the flow of liquid into waterways or sewers or offsite. Prevent discharge in all cases into storm or sewer drains by sealing off with plastic and/or containment dikes.

If the release has not been contained entirely on company premises or if the release is beyond the capability of the CRT, notify the contracted spill responder immediately. Additionally, regulations require the notification of the National Response Center at (800) 424-8802, as well as the appropriate local authorities.

After any type of incident, all emergency response equipment must be cleaned or replaced so that it is ready for its intended use. All waste must be packaged, labeled, and handled as hazardous waste. An Accident/Incident Report must be completed and given to the EH&S Department for identification of any possible corrective actions.

APPENDIX 'X-3' — TRANSPORTATION CHEMICAL SPILLS

For accidental releases, including hazardous materials transportation emergencies, we have selected CHEMTREC to handle our calls. CHEMTREC is a public service established by the Chemical Manufacturers Association (CMA). Although it is sponsored by private industry, its function and capabilities are recognized by the U.S. DOT. A formal memorandum of understanding (MOU) was signed by the CMA and DOT in 1980 which officially recognizes CHEMTREC as the central emergency response service for incidents involving the transportation of hazardous waste.

They provide 24-hour service and can be reached through a toll-free emergency telephone number — 800/424-9300. Their primary source for responding is product Material Safety Data Sheets (MSDSs).

Part V

Checking and Corrective Action

16 Monitoring and Measurement

16.1 INTRODUCTION

In Appendix I, the entire ISO 14001 framework is mapped out and shows how all of the various elements of the standard interrelate. You will note that immediately downstream from Environmental Programs is the requirement to Monitor and Measure. What this means is that you should have a system in place to measure and monitor the actual performance of your environmental objectives and targets and the progress of your environmental programs. The result of your monitoring and measuring is reported to your management review committee. The results of this data will determine how well your programs are maintaining compliance with your policy.

16.2 THE ISO 14001 REQUIREMENTS

Element 4.5.1, *Monitoring and Measurement*, of the Environmental Management Standards states:

> The organization shall establish and maintain **documented procedures** to monitor and measure, on a **regular basis**, the **key characteristics of its operations** and activities that can have a **significant impact** on the environment. This shall include the recording of information to **track performance, relevant operational controls** and **conformance** with the organization's environmental objectives and targets.
>
> Monitoring equipment shall be **calibrated** and maintained and records of this process shall be retained according to the organization's procedures.
>
> The organization shall establish and maintain a documented procedure for **periodically evaluating compliance** with relevant environmental legislation and regulations.

There are several key points to keep in mind (highlighted) when implementing and integrating this requirement. The first is there must be a documented procedure that will define how, who, what and when. The second was alluded to in the introduction and that is the requirement to use the system to track performance, controls and performance. The third requirement is that all environmental equipment used to monitor and measure must be calibrated in order to ensure the accuracy of the data. The fourth requirement is related to Element 4.3.2 in that you must monitor compliance to environmental regulations.

16.3   RELATIONSHIP TO ISO 9001

The primary section in ISO 9001 relating to the ISO 14001 requirements is found in 4.20, *Statistical Techniques*. Specifically, ISO 9001 states:

> **4.20.1, Identification of** Need   The supplier shall identify the need for statistical techniques required for establishing, controlling and verifying process capability and product characteristics.

> **4.20.2, Procedures**   The supplier shall establish and maintain documented procedures to implement and control the application of the statistical techniques identified in 4.20.1.

It has been my experience, however, that the integration of the ISO 14001 requirements into the ISO 9001 procedure was very difficult. The primary reason for this is that Element 4.20, *Statistical Techniques*, requires an "identification of need" that is not based on whether or not a process is "significant." Unless you also become familiar with the guideline document, ISO 9004, you may miss where the connection with ISO 14001 occurs. ISO 9004, the guideline document, states in Section 10 the following:

> **Section 10.1.2**   Verification of the quality status of a ... process, ... processed material, service, or environment should be considered at important points in the production sequence to minimize effects of errors and to maximize yields.

> **Section 10.1.3**   Monitoring and control of processes should relate directly to finished product specifications or to an internal requirement, as appropriate ... In all cases, relationships between in-process controls, their specifications and final product specifications should be developed ...

Thus, the ability to integrate ISO 14001 with ISO 9001 can only be accomplished if you fully understand the guidelines provided in ISO 9004 — your ISO 9001 procedure(s) should reflect ISO 9004 information.

16.4   A MONITORING AND MEASURING PROCEDURE

The primary key to complying with the ISO 14001 requirements is to establish a documented procedure that establishes *what* you are going to do, *how* you plan to do it, *why* you are doing it and *when* it will be done. If you do not have an adequate procedure within the ISO 9001 framework to build upon, I would suggest that you write another procedure to satisfy just the ISO 14001 requirements and then work at a later date on developing an integrated procedure. Although the intent of this book is to show you examples of how integration can be done, there are instances where the initial efforts may be much more difficult than anticipated. This may be due to the fact that there isn't a specific ISO 9001 procedure for monitoring and measurement. As you can see from Table 16.1, the correlation of the ISO 14001 requirements occurs in several different areas of ISO 9001 and, since ISO 14001 requires a documented procedure, integration may be difficult at first. This was my

TABLE 16.1
Correlation of "Monitoring and Measurement Requirements"

ISO 9001		ISO 9004		ISO 14001	
Section	Requirement	Section	Requirement	Section	Requirement
4.20.2	Documented procedure consistent with the requirement;	10.1.4	Documented test and inspection procedures should be maintained for each characteristic to be checked;	4.5.1(i)	Documented procedure to monitor and measure
4.20.2	Documented procedure to implement and control application of statistical techniques.	11.4	Processes that are important to product quality should be planned, approved, monitored and controlled.		
4.11.2 (a) 4.20.1	Define measurements to be made; Need for statistical techniques for establishing, controlling and verifying process capability and final product characteristics.	10.2	Operations associated with product or process characteristics that can have a significant effect on product quality should be identified.	4.5.1(i)	Monitor key characteristics that can have a significant impact
4.2.3 (f)	Identification of suitable verification at appropriate stages.	10.1.2	Verification of the quality status at important points in the production sequence to minimize effects of errors.	4.5.1 (i)	Track performance
4.9 (d)	Monitoring and control of suitable process parameters	10.1.4	All in-process and final verifications should be planned and specified.	4.5.1 (i)	Track relevant operational controls
4.10.4	Specified inspection and testing results meet specified requirements	10.1.3	Monitoring and control of processes should relate directly to finished product specifications	4.5.1 (i)	Track conformance with objectives and targets
4.11.2 (a) 4.11.2 (b)	Select appropriate measuring equipment; Calibrate monitoring equipment	13.2	Procedure for the control of inspection, measuring and test equipment	4.5.1 (ii)	Calibrate monitoring equipment
4.9 (c)	Compliance with reference standards/codes	19 (a) 19 (b) 19 (d)	Identify relevant safety standards; Carrying out design evaluation tests and prototype (or model) testing for safety; Developing a means of traceability to facilitate product recall.	4.5.1 (iii)	Documented procedure to evaluate environmental legislation compliance.

* (i) denotes paragraph
** (a) denotes subsection of element

own experience — too much time and effort was needed to bring the two standards together. With an ISO 14001 audit looming on the near horizon, I could not put too much effort into figuring out how to work with the ISO 9001 documents. The result was an EH&S-specific procedure for monitoring and measurement.

In this section, I have provided a rough draft of an EH&S-specific monitoring and measurement procedure based on the procedure template I have been using in several of the previous chapters. The referenced document numbers, of course, are made up. You will note that I have also included the "reporting" requirements found under Element 4.4.1 for the management representative. In addition, I would like to take this particular opportunity to show that all of the procedures I have provided need not be concerned with just environmental issues, but can also include your company's health and safety requirements. If you or your department has responsibility for health and safety as well, it makes sense for you to write your procedures to cover environmental, health and safety requirements. If international health and safety standards are ever developed (see Chapter 21), you would have done a lot of preparation beforehand. So with this in mind, let's take a look at an EH&S-specific procedure.

Purpose This procedure is intended to identify and document the monitoring, measurement and reporting requirements for the aspects and/or impacts of the company's environmental, health and safety programs.

Scope This procedure applies to all of the company's environmental, health and safety programs.

Definition of terms

- **ARB** Air Resources Board
- **VOC** Volatile Organic Compound
- **POC** Precursor Organic Compound
- **EPA** Environmental Protection Agency
- **OSHA** Occupational Safety and Health Administration
- **RCRA** Resource Conservation Recovery Act which regulates all hazardous waste from "cradle to grave."
- **EH&S Aspects** those elements of the business' activities, products and services which can interact or impact with the environment, people and property.
- **Environmental, Health and Safety Committee** The organizational body within the company which has the responsibility for reviewing the status of the company's environmental, health and safety programs on a periodic basis (need to specifically identify) and making recommendations/suggestions for corrective actions and improvements.
- **TWA** Time Weighted Average

Reference documents

Company Documents

- QAP 12345, Corrective and Preventive Action Procedure
- QAP 67891, Management Review Procedure

- QAP 13579, Calibration Control of Measuring and Test Equipment
- SOP 24681, EH&S Controlled Documents
- SOP 13963, Waste Minimization and Pollution Prevention Program
- SOP 14843, Air Sampling

Regulatory

- 22 CCR Chapter 1, Article 1, Waste Management Act
- 22 CCR Section 67750, Hazardous Waste Source Reduction and Management Review Act
- ARB Rules and Regulations 1-543, 2-1-403 and 1-404

International Standards and Practices

- ISO 14001, Element 4.3.1, Environmental Aspects
- ISO 14001, Element 4.5.1, Monitoring and Measurement
- ISO 9001, Element 4.11, Control of Inspection, Measuring and Test Equipment
- ISO 14001, Element 4.4.1, Structure and Responsibility

EH&S precautions

- Personnel who are responsible for obtaining monitoring and measurement data in areas or at processes which pose actual or potential significant hazards should observe all personal protective equipment requirements.

Precedence This procedure shall only be superseded by specific and required conditions or data collection techniques as prescribed by local, state and/or federal regulations.

Responsibilities

Environmental, Health, and Safety Manager

- Approve all outside vendors who have been contracted to conduct monitoring and measurement of an environmental, health or safety subsystem.
- Approve all EPA approved analytical laboratories including test methods, equipment requirements, etc.
- Provide final approval of any reports and/or data submitted by an outside contractor.
- Submit reports of monitoring and measurement data to appropriate local, state or federal regulatory agencies as required.
- Communicate the monitoring and measurement requirements to appropriate personnel.
- Assist in establishing the systems, procedures and forms to collect or assemble the monitoring and measurement data.
- On a periodic basis (as prescribed in the *procedure* section) collect the data.
- As prescribed either by regulatory requirements or internal procedures, report the data results to appropriate personnel. The reports may be used:
  — As an update of the status of department-specific objectives and targets;
  — For submittal for regulatory compliance; and/or

— By the Management EHS Committee as an update of the status of the company's overall objectives and targets. This review may result in a change in some objectives and targets and potential corrective actions stemming from nonconformances.

- Distribute annual reports to company senior management for review and evaluation for the potential purposes of changing the Environmental, Health and Safety Policy, standards and guidelines, procedures and/or the establishment of new annual objectives and targets.

Company Employees

- Ensure all logs, data collection forms, or computerized data collection systems are completed as required by the process.

Department Management

- Ensure personnel who are direct reporters understand and complete the requirements.

Tools   N/A

Materials   N/A

Procedure

- A. Flowchart (Appendix A)
- B. Reporting Schedule (Appendix B)
- C. Environmental, Health and Safety Manager
  1. Determines if the monitoring, measurement and/or reporting is per regulatory requirements or per other (company, divisional, site, corporate, industry standard, etc.)
- D. EPA Requirements
  1. Air Emissions (VOC and POC) shall be monitored on a monthly basis. Summary data is obtained by adding up the total quantity of VOC or POC chemical used during the previous month. The amount of VOC or POC chemical used over the previous 12 months shall not exceed the maximum quantity allowed under the CONDITIONS of the Permit-to-Operate issued by the ARB. Monthly reports are distributed for review to appropriate departments, management, and the EH&S Committee.
  2. Liquid and Solid Hazardous Waste is tracked using the RCRA-approved Hazardous Waste Manifest. The manifest is completed whenever a hazardous waste is prepared and transferred from the facility. The quantifiable data on the form is logged into the EH&S department's hazardous waste database. Monthly reports of department-specific waste generation are distributed for review and comparison against department goals and objectives. A summary report is reviewed on a monthly basis by the EH&S Committee.
- E. OSHA Requirements
  1. Injuries and Illnesses are recorded as each event occurs. At the end of the second week of each month, the Worker Protection group distributes

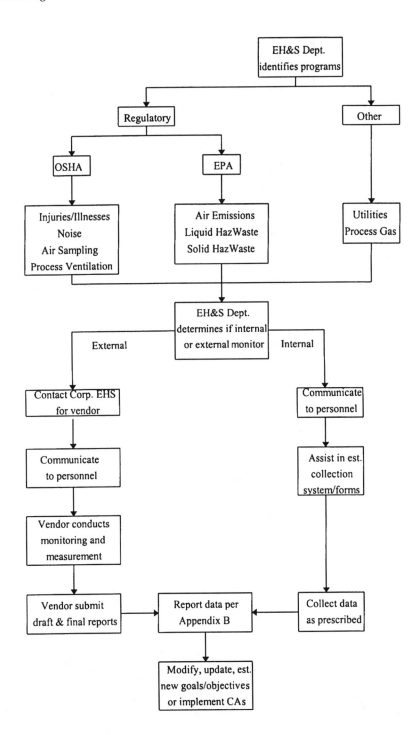

APPENDIX A.   Procedure flowchart.

APPENDIX B
Reporting Schedule Table

Data/Metric	Monthly				Quarterly				Annually			
	Dept. Team	Corp. EH&S	Senior Mgmt.	EH&S Comm.	Dept. Team	Corp. EH&S	Senior Mgmt.	EH&S Comm.	Dept. Team	Corp. EH&S	Senior Mgmt.	EH&S Comm.
1. Air Emissions	•		•	•	•	•	•	•	•	•	•	•
2. Liquid HazWaste	•		•	•	•		•	•	•		•	•
3. Solid HazWaste	•		•	•	•		•	•	•		•	•
4. Injuries/Illnesses	•		•	•	•	•	•	•	•			•
5. Noise Surveys	•	•	•	•				•				•
6. Air Sampling		•	•	•				•				•
7. Process Ventilation		•		•				•				•
8. Utilities				•				•			•	•
9. Process Gas				•				•			•	•
10. Recycled Material				•				•			•	•

the OSHA 200 log. Data from the log is analyzed to update department-specific and overall company metrics. Monthly reports are distributed per the schedule in Appendix B.

2. Noise Surveys are contracted out to an approved vendor. Area and personal dosimetry monitoring is required to provide measurement of a person's exposure to noise based on an 8-hour Time Weighted Average (TWA). Area monitoring is also conducted to assess community noise concerns from the facility. The draft and final reports are evaluated by EH&S with results distributed to appropriate personnel and management. Specific measurements may be used to implement control measures (engineering, administrative and/or personal protective equipment).

3. Air Sampling is conducted per the requirements detailed in SOP 14843.

4. Process Ventilation monitoring is conducted by an outside vendor. Measurements for flow rate, face velocity and/or capture velocity are conducted and subsequently posted on the source. The final report is reviewed by EH&S to assess and implement control measures for any part of the system which is deficient.

F. Other Requirements

1. Utility Consumption data for power, water and natural gas is monitored by the local utility agencies. Results for the company are consolidated on a monthly basis by the Facilities Department and copied to the EH&S Department. Data analysis results are reported to the EH&S Committee on a monthly basis showing current performance against objectives and targets.

2. Processes Gas consumption is monitored by an in-line meter before entering the building and the total monthly volume is calculated. Data analysis results are reported to the EH&S Committee on a monthly basis showing current performance against objectives and targets.

3. The Recycling Program involves the separation of previously identified solid waste streams. The EH&S Department weighs each recyclable waste stream as it is transferred to an approved recycling vendor. The data is accumulated for analysis and is reported to the EH&S Committee on a monthly basis.

G. The EH&S Department determines if the monitoring of programs is to be done internally or externally (e.g., vendor or a corporate group).

H. External Monitoring and Measurement

1. The EH&S Department contacts either the vendor or the appropriate corporate group.

2. The EH&S Department communicates with appropriate personnel or departments regarding the monitoring and measurement requirements.

3. The vendor conducts the monitoring and measurement.

4. The vendor submits a draft report to the EH&S Department.

5. Final draft is approved.

6. Report data is communicated to appropriate personnel (see Appendix B).

I. Internal Monitoring and Measurement
1. The EH&S Department communicates with appropriate personnel or departments the monitoring and measurement requirements.
2. The EH&S Department assists in establishing the data collection system and collects the data.
3. Report data is communicated to appropriate personnel (see Appendix B).
J. Department management and/or EH&S Committee modifies, updates or establishes new goals and objectives or implements corrective actions.

16.5 RECORDING OF INFORMATION

Now that you have a procedure which defines the ISO 14001 requirements, the next step involves determining what kind of data you need to record and collect for reporting. When doing this, however, it is important to remember one of the basic requirements of Element 4.5.1:

> This shall include the recording of information to track performance, relevant operational controls and conformance with the organization's environmental objectives and targets.

In other words, don't spend your time recording and tracking data that will not provide you with any significant information and value — your data should relate back to your objectives and targets.

In the next few sections I will provide some examples of spreadsheets and graphs which you might use as templates for setting up your own program. I will be focusing strictly on the recording and collection of environmental data in the areas of air emissions, hazardous waste, recycling, natural resources, and other community impacts.

16.5.1 AIR EMISSIONS

Whether your company is located in an *attainment* or *nonattainment* area by the Air Resources Board and the EPA, the chemicals you emit into the atmosphere may perhaps have the greatest tendency to draw attention from your neighbors and the local regulatory agencies. This is especially true due to the large amount of media attention being given to the chemicals and industries creating global warming. If your company emits some form of vapor or gaseous substance (whether harmful or not), the local public and nongovernmental organizations (NGOs) will be paying much more attention to you. Thus having objectives and targets to minimize your emissions may be based solely on the influence of "other interested parties" and not because of any regulatory requirements — your objective may be to maintain a good relationship with your neighbors.

In this first example, I am going to assume that you have a particular operation that requires a permit-to-operate from the local Air Resources Board. In addition, the permit-to-operate has requirements or *conditions* which must be met and that the *condition* is not to exceed the use of a given quantity of a Volatile Organic

TABLE 16.2
Air Emissions Data for Source #246

Month	Qty. (gal)	Liquid Waste	Net Usage	Total Usage	% of Max.	12-Month Projection
January	25	2	23	23	3.83	276
February	30	5	25	48	8.00	288
March	20	1	19	67	11.17	268
April	45	8	37	104	17.33	312
May	37	4	33	137	22.83	329
June	43	6	37	174	29.00	348
July	33	3	30	204	34.00	350
August	62	4	58	262	43.67	393
September	55	2	53	315	52.50	420
October	47	4	43	358	59.67	430
November	56	3	53	411	68.50	448
December	49	5	44	455	75.83	455
Maximum				600		

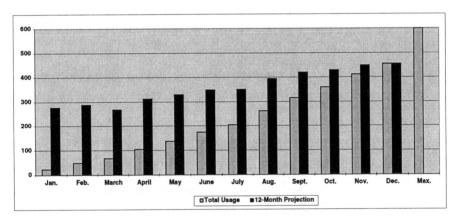

CHART 1. Air emissions data for source #246.

Chemical (VOC) during any consecutive 12-month cycle. What needs to be done then is to develop a database which will assist you in measuring your performance against this condition and against the particular objective and target associated with this process. A spreadsheet collecting this data may look like Table 16.2 and Chart 1.

The last column is used primarily for new permits and shows management how well the operating projections match up against the new permit conditions. This will allow you to make adjustments in the permit before noncompliance can occur.

In this particular case, the company's objective is to stay below the maximum allowable consumption of chemical with the target being <600 gallons in any 12-month cycle. Both the objective and target also happens to be a regulatory requirement. This particular database, however, may be one of several air emission sources such that the total tonnage of VOCs may be considerable. If that is the case, then other issues or objectives must be evaluated and the data used to verify performance against them. Other objectives based on legal, technological, financial, operational and external considerations may include:

- Meeting requirements of a Significant Minor Operating Permit (SMOP) under Title V of the Clean Air Act;
- Eliminating the need for Abatement Control Technology;
- Reducing material and operating costs;
- Developing a closed-loop system to recover volatile emissions and eliminate "visible" discharges; and/or
- Marketing and sales advantages.

16.5.2   HAZARDOUS WASTE MINIMIZATION

As we discussed in Chapter 5, one of the primary objectives of your environmental management program should be the "prevention of pollution." Time, in fact, was spent in discussing the difference of this requirement with "pollution prevention" which the U.S. EPA wanted to have as a requirement in the ISO 14001 Standards. Although I will not be discussing this further, the distinction can have a tremendous impact on your objectives, targets and programs. For instance, if two companies or even two different departments in the same company have identical processes and waste streams, management objectives may be different and what each ends up monitoring and measuring will be affected by the decision.

In many countries environmental legislation has been focusing on the reduction of hazardous waste at the source (i.e., "pollution prevention") rather than focusing on what to do with the waste once it has been generated (i.e., "prevention of pollution"). In California, for instance, environmental legislation includes the Hazardous Waste Source Reduction and Management Review Act (SB 14) which requires the development of plans and programs that will reduce the amount of hazardous waste being generated at a particular source (i.e., a manufacturing process step). Although financial resources for many companies may prohibit source reduction programs, it is my belief that "pollution prevention" should be management's focus and not directing resources towards figuring out what to do with the waste after it has been generated. With that in mind the databases in the rest of this section will focus on "pollution prevention" efforts and the next section will focus on "prevention of pollution" and recycling efforts.

The monitoring and measuring of liquid and solid hazardous waste should again focus on those processes which have an actual impact or the potential to impact the environment. Let's look at a process which generates methanol (liquid) and methanol contaminated debris (solid) waste.

Let's assume that during the previous calendar year (1996) your process generated 100 gallons of methanol waste and annual production was 275,000 widgets. For 1997, you realize that production will go up ~10% to 300,000 widgets, but you would like to establish an objective to reduce methanol waste to 85 gallons. A good way to track performance is to do it in terms of a rate–volume of waste per production units. In this

TABLE 16.3
1997 Methanol Waste Generation

Month	Waste (oz.)	Waste Sum (oz.)	Parts Produced (Qty./1000)	Total Parts (Qty/1000)	Rate
January	640	640	9,300	9,300	0.069
February	640	1,280	22,000	31,300	0.041
March	1280	2,560	25,375	56,675	0.045
April	1280	3,840	28,475	85,150	0.045
May	1920	5,760	23,660	108,810	0.053
June	2560	8,320	27,668	136,478	0.061
July	1280	9,600	31,005	167,483	0.057
August	640	10,240	26,555	194,038	0.053
Sepember	0	10,240	27,845	221,883	0.046
October	640	10,880	29,457	251,340	0.043
November	0	10,880	25,893	277,233	0.039
December	640	11,520	28,765	305,998	0.038
Target		12,000		300,000	0.04
1996		12,800		275,000	0.047

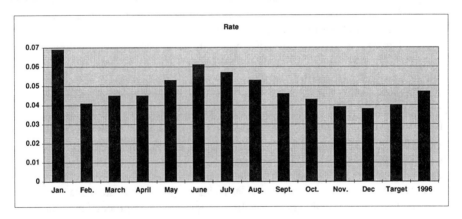

CHART 2. 1997 methanol waste generation.

case we have ounces of methanol waste per 1000 widgets produced. For solid waste, you can set a rate of pounds or kilograms per quantity of product manufactured.

The reason for using rates is that it eliminates business fluctuations and makes it easier to track performance from year to year rather than on a total volume of waste generated. Table 16.3 and Chart 2 show the 1997 data and also includes the results of 1996 and the target rate. By including this data you can chart your performance each month and include your target and last year's results on it for comparison. As you can see from this particular table, the data shows that the objective was successful.

Monitoring the progress of the target in this manner will assist you in making any "course corrections." You will note that through June, there was a steady increase in the rate that exceeded the target. This, of course, warranted a review of the objective and an examination of the process to determine why the volume of waste was increasing. During the second half of the year, the volume of waste decreased and two objectives were met: the overall waste generation and the rate. The intent is to monitor the data often enough to make changes in time to meet the objective(s) and target(s) at the end of the year.

16.5.3 Solid Waste and Recycling

Hazardous waste has had the attention of the public and the regulatory agencies for about two decades. Recently, however, agencies on the federal, state and local levels have been putting an ever-increasing focus on nonhazardous waste. With diminishing natural resources and landfill space, the local public and environmental "watch dog" organizations have been pressuring city and county governments to decrease the amount of trash and garbage going into the local municipal landfill. This is especially true for larger cities as building construction takes up more and more of the open spaces. Landfills end up being far removed from the waste collection companies.

The response from state and federal governments has been to pass legislation that requires local agencies and governments to implement waste reduction and recycling programs. In California, for instance, state government passed the California Integrated Waste Management Act that required cities and counties to reduce solid waste at least 25% by 1995 and to reduce 50% by the year 2000. Enforcement of such legislation affects residential, commercial, industrial and institutionalized government waste. The solution almost everyone has implemented has been to initiate a recycling program.

Although residential waste accounts for the largest percentage of landfill and recycled material, large corporations draw particular attention from local government. Exclusive of this attention, the reduction of solid waste makes good financial sense and contributes to sustainable development (Chapter 2). Independently of this, if we look at solid waste purely from an ISO 14001 standpoint, in most cases this aspect of your corporation will turn out to have a significant impact on the environment.

Any monitoring and measurement associated with your solid waste should be done in conjunction with your recycling program and in-process manufacuring yields. As with the above sections, metrics can use a rate (tons/production) for the purposes of the monitoring. Table 16.4 and Chart 3 show an example of a database used to monitor and measure solid waste. Column 1 is the total waste generated per month and column 2 is the running total. If we also include the amount of product made (times 1 million) from Table 16.3, then we can calculate a rate. The chart includes the previous year's results and the current year's target as a means to track performance. As with the hazardous waste metrics, this format allows you to notice that your waste generation during the first six months is exceeding your target. This information can assist management in evaluating any potential causes for the increase in solid waste and making an adjustment to get operations back on target. As with the two previous tables and charts, this information allows you to track performance

TABLE 16.4
1997 Solid Waste Generation

Month	Tons	Sum Total	Total Cost ($25/ton)	Production (× M)	Total Prod. (× M)	Monthly Rate	Annual Rate
January	48	48	1,200	9,300	9,300	0.0052	0.0052
February	60	108	2,700	22,000	31,300	0.0027	0.0035
March	55	163	4,075	25,375	56,675	0.0022	0.0029
April	57	220	5,500	28,475	85,150	0.002	0.0026
May	61	281	7,025	23,660	108,810	0.0026	0.0026
June	63	344	8,600	27,668	136,478	0.0023	0.0026
July	65	409	10,225	31,005	167,483	0.0021	0.0024
August	69	478	11,950	26,555	194,038	0.0026	0.0025
September	65	543	13,575	27,845	221,883	0.0023	0.0024
October	63	606	15,150	29,457	251,340	0.0021	0.0024
November	60	666	16,650	25,893	277,233	0.0023	0.0024
December	55	721	18,025	28,765	305,998	0.0019	0.0024
Target		750	18,750		300,000		0.0025
1996		800	20,000		275,000		0.0029

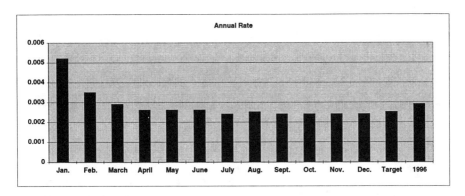

CHART 3. 1997 solid waste generation.

against your objectives and targets and to comply with the commitment to "prevention of pollution" in your environmental policy.

In conjunction with the data in Table 16.4, you will most likely be keeping track of your recycling efforts. Implementing a program to recycle white and mixed paper, cardboard, metal, newspapers, plastic, glass and aluminum cans is a common practice these days. Recycling metrics, however, are rarely utilized as a tool to measure performance against a prevention of pollution program nor recognized as being an actual or potential significant aspect of your operation.

Table 16.5 and Chart 4 shows data for a typical recycling program, but by itself it does not really provide any information to assist management in evaluating performance

TABLE 16.5
Recycling Metrics (tons)

Month	White Paper	Cardboard	Plastic	Metal	Glass	Newspapers	Mixed Paper	Monthly Total
January	1.1	2.1	0.11	20.3	0.15	2.5	0.01	26.27
February	1.2	2.2	0.21	19.4	0.12	2.8	0.02	25.95
March	0.9	2.4	0.22	18.5	0.15	2.4	0.04	24.61
April	0.6	2	0.14	17.6	0.08	2.6	0.01	23.03
May	1.3	1.9	0.15	18.3	0.09	2.7	0.02	24.46
June	1	1.7	0.14	19.6	0.11	3	0.03	25.58
July	1.2	1.9	0.1	18.5	0.12	3.2	0.05	25.07
August	1.3	2	0.15	19.7	0.14	3.3	0.06	26.65
September	1.4	2.6	0.18	20.3	0.11	3.2	0.05	27.84
October	1.5	2.5	0.15	21.4	0.12	3.1	0.07	28.84
November	1.6	2.7	0.18	22	0.16	3.1	0.05	29.79
December	1.5	2.5	0.19	22.3	0.18	3.5	0.06	30.23
Total	14.6	26.5	1.92	237.9	1.53	35.4	0.47	318.32

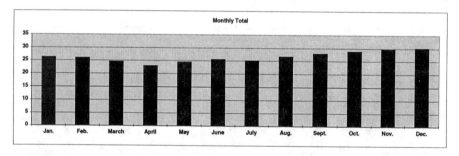

CHART 4.   Recycling metrics (tons).

against a prevention of pollution program. The key is to compare it and use it in conjunction with the solid waste metrics. The information is expanded in Table 16.6 and Chart 5 to show the total recyled and its net impact.

16.5.4   UTILITIES AND NATURAL RESOURCES

Whether residential, commercial or industrial, the consumption of electrical power, natural gas, coal fuel and water can have a profound effect on your finances. For larger firms a monthly utility bill can reach hundreds of thousands of dollars per month and millions of dollars over a period of several years. Financial considerations are always at the forefront of a manager's mind in this case, but in terms of ISO 14001, natural resource consumption should be considered as an actual or potential significant impact on the environment. It should also be considered as an important aspect of a sustainable development program.

TABLE 16.6
Waste and Recycling Metrics

Month	White Paper	Cardboard	Plastic	Metal	Glass	Newspapers	Mixed Paper	Total Recycled	Solid Waste	Hazardous Waste	Air Emissions	Net Waste	Net Impact
January	1.1	2.1	0.11	20.3	0.15	2.5	0.01	26.27	48	0.0297	0.106	48.1357	21.8657
February	1.2	2.2	0.21	19.4	0.12	2.8	0.02	25.95	60	0.0425	0.128	60.1705	34.2205
March	0.9	2.4	0.22	18.5	0.15	2.4	0.04	24.61	55	0.0468	0.085	55.1318	30.5218
April	0.6	2	0.14	17.6	0.08	2.6	0.01	23.03	57	0.0765	0.191	57.2675	34.2375
May	1.3	1.9	0.15	18.3	0.09	2.7	0.02	24.46	61	0.0808	0.157	61.2378	36.7778
June	1	1.7	0.14	19.6	0.11	3	0.03	25.58	63	0.1105	0.183	63.2935	37.7135
July	1.2	1.9	0.1	18.5	0.12	3.2	0.05	25.07	65	0.0553	0.14	65.1953	40.1253
August	1.3	2	0.15	19.7	0.14	3.3	0.06	26.65	69	0.0383	0.264	69.3023	42.6523
September	1.4	2.6	0.18	20.3	0.11	3.2	0.05	27.84	65	0.0085	0.234	65.2425	37.4025
October	1.5	2.5	0.15	21.4	0.12	3.1	0.07	28.84	63	0.0383	0.2	63.2383	34.3983
November	1.6	2.7	0.18	22	0.16	3.1	0.05	29.79	60	0.0128	0.28	60.2928	30.5028
December	1.5	2.5	0.19	22.3	0.18	3.5	0.06	30.23	55	0.0085	0.208	55.2165	24.9865
Total	14.6	26.5	1.92	237.9	1.53	35.4	0.47	318.32	721	0.5485	2.176	723.7245	405.4045

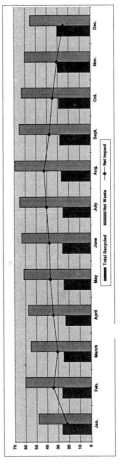

CHART 5. Waste and recycling metrics.

TABLE 16.7
Annual Power Consumption Comparison (KWH × 1000)

Month	1992 Power	1993 Power	1994 Power	1995 Power	1996 Power	1997 Power
January	645	786	633	670	735	890
February	730	745	640	616	777	1006
March	855	942	664	741	715	968
April	900	896	609	714	701	927
May	1055	899	658	555	845	977
June	932	989	832	526	912	1009
July	1065	971	757	650	901	1123
August	885	910	600	724	967	1178
September	1062	956	850	759	1004	1157
October	906	814	714	678	960	989
November	793	592	644	657	818	957
December	815	693	652	653	924	875
Average	887	849	688	662	855	1005

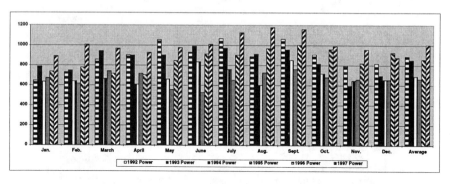

CHART 6. Annual power consumption comparison (KWH × 1000)

Table 16.7 and Chart 6 show some typical metrics for monitoring power consumption. As for most companies, the high usage in winter and summer months is most likely due to heating and air conditioning increases. In this scenario we'll assume that the increase in 1996 and 1997 was due to an increase in sales and a potential increase in work hours. This data, however, only shows the monthly and annual consumption in comparison to previous years and does not take into consideration activities associated with fluctuations in production.

If we look at Table 16.8 and Chart 7, however, you will see the data in terms of a rate (KWH/production) as a means of assessing how efficiently a company is using its power. If sales continue to increase over the next few years, it is to be expected that power consumption will need to increase accordingly. If total consumption is your only way of measuring, then reduction criteria under ISO 14001

TABLE 16.8
Power Consumption Rate

Month	KWH	Total KWH	Production (× 1000)	Total Production	Rate
January	890,312	890,312	42.50	42.50	20,949
Febreuary	1,005,647	1,895,959	62.70	105.20	18,022
March	968,136	2,864,095	44.21	149.41	19,169
April	926,735	3,790,830	49.92	199.33	19,018
May	976,548	4,767,378	43.61	242.94	19,624
June	1,008,942	5,776,320	45.84	288.78	20,002
July	1,123,444	6,899,764	50.60	339.38	20,330
August	1,178,248	8,078,012	38.50	377.88	21,377
September	1,156,947	9,234,959	55.00	432.88	21,334
October	989,427	10,224,386	61.60	494.48	20,677
November	956,746	11,181,132	56.48	550.96	20,294
December	874,948	12,056,080	44.76	595.72	20,238
1996		10,260,347		480.02	21,375

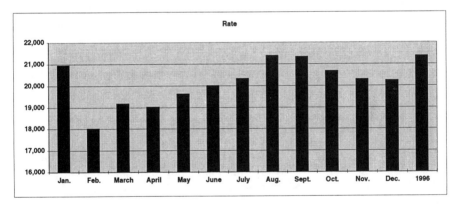

CHART 7. Power consumption rate.

may be very difficult to meet. By initiating programs to reduce consumption based on the rate, rather than the overall consumption, you can provide an auditor with enough evidence to show you are conserving natural resources.

The data tables and charts presented for power can be used to monitor and measure the other types of natural resources as well. Water consumption rates (gallons per production), natural gas rates (cubic feet per production), coal fuel (tons or pounds per production) and other process gases (cubic feet per production) can be established and utilized to drive overall efficiency in their use. As with all of these natural resources, the data needs to be used to drive continual improvement and sustainable development programs. Most management typically responds by communicating to everyone to turn off lights and equipment when not in use, but an energy conservation program needs to go beyond this. Contracting for an energy survey will go a long way towards identifying specific equipment and/or areas wasting energy.

16.6   CALIBRATION

If your organization is certified to ISO 9001, you most likely have a very extensive and well-organized calibration program due to the extensive requirements which ISO 9001 has in this area. I do not plan to spend too much time discussing this subject, but I do want to make two points.

First, ISO 9001 requires in the control procedure (Section 4.11.2) that the supplier shall *"identify all inspection, measuring and test equipment that can affect product quality, and calibrate and adjust them at prescribed intervals, or prior to use, against certified equipment having a known valid relationship to internationally or nationally recognized standards."* The procedure in most cases will contain a list of equipment which fall under the requirements. If that is the case, it is a very simple matter to include any environmental equipment used for monitoring and measurement to be included on this list.

Second, if you utilize an outside contractor or vendor to conduct monitoring and measurement of an aspect of your operation which has an actual or potential impact on the environment, then you should be requesting information regarding the calibration of their equipment. You must be assured that the vendor's equipment is going to provide you with accurate information and data. If their information is inaccurate, it could have consequences and impact your ability to meet some specific objectives and targets.

16.7   WHAT AUDITORS WILL LOOK FOR

The first thing an auditor will ask for is a documented procedure which defines the how, when, what, who and why of the monitoring and measurement program. Second, they will assess whether or not the results of your monitoring and measurement are being reviewed by appropriate personnel. Third, an auditor will want to connect the results with specific objectives and targets. Fourth, an auditor will verify if the results indicate continual improvement of a process and whether or not it is contributing to the "prevention of pollution."

17 Nonconformance and Corrective and Preventive Action

17.1 INTRODUCTION

An effective corrective action and preventive action program is one of the foundational building blocks for any system, including Quality and Environmental Management. It is not good enough to put any system in place without having a process to consistently improve the system and to address any nonconformances which may arise from such sources as:

- internal or external inspections.
- internal and external audits.
- customer inquiries and questionnaires.
- management reviews.
- customer complaints.
- internal complaints.
- training deficiencies.
- emergency response improvements.
- regulatory noncompliance.

You will find that the integration of the two standards can be accomplished quite easily when it comes to the procedure and the reporting forms being used. An issue may arise, however, when it comes to assigning who will have responsibility for addressing the corrective actions themselves. In most organizations, quality-based corrective actions may be addressed by a management team that may meet only once a week. For environmental corrective actions this may not be adequate — potential legal issues and other liabilities will warrant a much quicker response time. What ends up happening as a result is that the QA management team does not want to assume responsibility for environmental corrective actions. All too often, the environmental department is told to "take care of it" without any other management involvement.

17.2 COMPARISON OF THE STANDARDS

Table 17.1 shows the side-by-side comparison of the two standards. You will immediately notice that the verbiage is almost identical. As with some of the other ISO 14001 requirements, the drafters used the ISO 9001 requirements as a template.

You may have noticed that the ISO 14001 standard does not specifically address preventive action to the detail that ISO 9001 does. However, it should not be assumed that an auditor will give cursory attention to this requirement. Although a preventive action system is important, a corrective action system will command much more

TABLE 17.1

Correlation of "Corrective and Preventive Action" Requirements

ISO 9001		ISO 14001	
Section	**Requirement**	**Section**	**Requirement**
The supplier shall establish and maintain documented procedures for ...		*The organization shall establish and maintain procedures for defining responsibility and authority for...*	
4.14.2 (a)	Handling customer complaints reports of product nonconformities	4.5.2 (i)	Handling and investigating nonconformance
4.14.2 (b)	Investigation of the cause of nonconformities relating to product, process and quality system, and recording the results of the investigation		
4.14.2 (c)	Determination of corrective action needed to eliminate the cause of nonconformities	4.5.2 (i)	Taking action to mitigate any impacts caused
4.14.3 (b)	Determination of steps to deal with problems requiring preventive action		
4.14.1 (i)	Implementing corrective and preventive action	4.5.2(i)	Initiating and completing corrective and preventive action.
4.14.2 (d)	Control applications ensuring corrective action is taken and it is effective		
4.14.3 (c)	Initiating preventive action and controls to ensure effectiveness		
Any corrective or preventive action taken to eliminate the causes of actual or potential nonconformities shall be ...		*Any corrective or preventive action taken to eliminate the causes of actual and potential nonconformances shall be ...*	
4.14.1 (ii)	To a degree appropriate to the magnitude of problems and commensurate with the risks encountered	4.5.2 (ii)	Appropriate to the magnitude of problems and commensurate with the environmental impact encountered
The supplier shall implement and record any...		*The organization shall implement and record any ...*	
4.14.1 (iii)	Changes to the documented procedures resulting from corrective and preventive action.	4.5.2 (iii)	Changes in the documented procedures resulting from corrective and preventive action.

attention from management and will be the primary focus of this chapter because of its potential impact on many other systems.

17.3 DEFINING A NONCONFORMANCE

Before we can implement a corrective and preventive action system, it is first important to define what a nonconformance is. Understanding this will assist in writing your procedure and help define the actions needing to be taken by a corrective action plan.

A nonconformance may be defined as "the failure to comply with some specified standard or criteria." In addition to broad categories listed in Section 17.1, a specified standard or criteria may include:

- Product quality
- Material quality
- Environmental monitoring and measuring requirements
- Operational or process controls
- Document controls
- Record retention controls

When defining a nonconformance, it is also important to understand the level of impact it may have on your environmental management system. ISO 9001 and ISO 14001 auditors typically define nonconformances as either *major* or *minor*. A major nonconformance will be one where there is an actual or potential deficiency that will seriously affect your quality or environmental management system. Some organizations call these *Hold Points* and are ones which carry sufficient weight to prevent certification to an ISO 9001 or ISO 14001 standard. This is typically caused by the absence of a major auditing criteria such as a documented procedure or failing to implement a procedure. A minor nonconformance is one where a deficiency will not affect the system or certification to the standard. However, this is not to say that the accumulation of a large number of deficiencies will not have the potential to create a major nonconformance. For instance, several minor nonconformances under ISO 14001 Element 4.5.2 may cause an auditor to consider them as equivalent to a major nonconformance for the whole element and end up classifying them as a Hold Point and, therefore, denying certification to the standard. Normally there is no limit to the number of minor nonconformances, but the professional judgment of the auditor must be considered.

17.4   CORRECTIVE ACTION

Once a nonconformance has been identified, the next step is to initiate the corrective action process. In the quality guideline document, ISO 9004, there is a detailed list and itemized description of what a corrective action program should include:

- Detection of a problem (Nonconformance)
- Assignment of responsibility ⎤
- Evaluation of importance
- Investigation of importance
- Investigation of possible causes ⎥ Corrective Action
- Analysis of problem
- Elimination of causes
- Process controls ⎦
- Permanent changes (Preventive Action)

In some organizations environmentally related corrective actions are not reviewed by a general management review board, but are directed towards the Environmental (Health and Safety) Committee for action. In the procedure I will present in the next section, I will make some notations regarding the responsibility of the Environmental Committee as they pertain to corrective actions.

17.4.1 CORRECTIVE ACTION PROCEDURE

With these criteria in mind, let's write a sample corrective action procedure which encompasses the requirements of ISO 9001 and ISO 14001. The template will utilize the main subject headings which I have been using in the previous chapters that are listed in Section 9.2.7.

Purpose   This procedure describes the process for initiating, recording, tracking and ensuring formal corrective actions stemming from identified nonconformances in areas affecting quality and environmental, health and safety activities.

Scope   This procedure applies throughout the company and addresses corrective actions stemming from all aspects of the business (quality, environmental, marketing, operations, material control, purchasing, etc.)

Definition of terms

- A **Minor Nonconformance** is one where a deficiency will not affect the quality or EH&S system or certification to a related standard.
- A **Major Nonconformance** is one where there is an actual or potential deficiency that will seriously affect your quality or environmental management system.
- **SPC**: Statistical Process Control
- A **Corrective Action Request** results from internal audits, customer complaints/audits, internal EH&S corrective actions and third party audits.
- An **Audit Corrective Action** results from a nonconformance found during a customer or third party (i.e., ISO, UL, CSA, EPA, OSHA, etc.) audit.
- A **Management Review Board** is responsible for classifying a nonconformance, assigning responsibility and completion timeframe requirements, and providing necessary resources to complete the corrective actions.
- The **Environmental Management Committee** is responsible for classifying an environmentally related nonconformance, assigning responsibility and completion timeframe requirements, and providing necessary resources to complete the corrective actions.
- A **Supplier Corrective Action** is a request for corrective action from a supplier as a result of quality defects on supplied materials and/or defects associated with service.

Referenced documents

- ISO 14001, Element 4.5.2, Nonconformances and Corrective and Preventive Action
- Title 8 California Code of Regulations, Section 3203(a)(6)

- ISO 9001, Element 4.14, Corrective and Preventive Action
- SOP #12345, Customer Complaints
- SOP #67891, Legal and Other Requirements
- SOP #24681, Management Review

EH&S precautions   NA

Precedence   If there is a conflict between this procedure and a customer requirement or regulatory agency, the latter shall take precedence.

Responsibilities

Quality Assurance Manager
- Maintaining and revising this procedure as needed.
- Ensuring documentation and records requirements are maintained.
- Tracking progress of a quality related corrective action.
- Facilitating and leading the Management Review Board in conjunction with the Environmental Manager (as appropriate). The board will be responsible for assigning the level of risk which includes a corresponding timeframe for completion:
  - Class 1   A critical nonconformance or issue which shows a major deficiency in a system and which may have the potential for jeopardizing a standards certification.
    To be completed within one week
  - Class 2   For multiple nonconformances or issues within an area which may or may not have the potential for jeopardizing a standards certification.
    To be completed within one month
  - Class 3   Any nonconformance or issue which is considered a minor deficiency in a process and warrants a process improvement.
    To be completed within three months

Department Management

- Identifying the basic cause of the nonconformance.
- Providing all necessary resources to complete the corrective actions in the timeframe specified by the Management Review Board.
- Reporting on status of corrective actions to the Quality Manager and/or Environmental Manager.

Environmental Manager

- Providing assistance to the Management Review Board for correcting an environmental nonconformance.
- Assigning a level of risk or Hazard Potential (HZPO) to the corrective action. The HZPO are classified by "class" which includes a corresponding timeframe for completion:
  - Class 1   Any condition or practice with potential for: (a) permanent disability, loss of life or body part; (b) extensive loss of structure/equipment or material; and/or (c) extensive environmental/ecological damage.

To be completed within 1–12 hours
— Class 2   Any condition or practice with potential for: (a) serious injury/illness; (b) property damage that is destructive but less severe than Class 1; and/or (c) environmental/ecological damage less severe than Class 1.
To be completed within 12–36 hours
— Class 3   Any condition or practice with potential for: (a) a nondisabling injury or illness; (b) nondisruptive property/equipment damage; and/or (c) nondestructive environmental/ecological damage.
To be completed within 36–72 hours
— Class 4   Any other condition or practice, such as those involved in engineering modifications with a corrective action time to be discussed and agreed upon by the concerned parties.
* Tracking progress of an environmentally related corrective action.
* Training personnel on all of the EMS requirements.

Procedure

* An internal *quality* audit corrective action is initiated by an internal audit report and is directed to the Management Reviews Board which assigns responsibility and level of risk. The specific nonconformances are noted on the corrective action form (Appendix A) and a cover letter is attached describing the system requirements (Appendix B). The responsible person must submit a corrective action plan with implementation dates within three (3) calendar days for Class 1 and ten (10) calendar days for Class 2 and 3 after receipt of the corrective action request.
* An internal *environmenal* audit corrective action is initiated by an internal audit report and is directed to the Environmental Management Committee which assigns responsibility and level of risk. The specific nonconformances are noted on the corrective action request form (Appendix A) and a cover letter is attached describing the system requirements (Appendix B). The responsible person must submit a corrective action plan with implementation dates within four (4) hours for Class 1, within twelve (12) hours for Class 2, within one (1) day for Class 3, and within one (1) week for Class 4 after receipt of the corrective action request.
* An *external audit* corrective action is at the discretion of the auditors for a nonconformance found during an external audit, except in cases involving regulatory agencies (EPA, OSHA, ARB, etc.) in which case the corrective action will be implemented by the Environmental Manager. The area assigned the corrective action will respond with a formal corrective action within the allotted time given.
* A *supplier* corrective action is initiated by Quality Control. Quality Control will evaluate the request and assess whether the nonconformance is quality in nature or environmental in nature and direct the corrective action request to the Management Review Board or to the Environmental Management Committee.

APPENDIX A

(Company Name)	**Corrective Action Request** **Proprietary Information**	**No.**

Originator of Request:	**Date:**	**Expected Completion:**

Description of Condition or Product Requiring Action (attach any other documents/reports):	Source of CAR and Report No. [ ] Discrepancy Report: _____ [ ] Audit Report: _____ [ ] Customer Return: _____ [ ] Customer Complaint: _____ [ ] Operations: _____ _____ [ ] Other Source: _____
Apparent Cause (and other possible causes):	Classification: [ ] Critical (Class 1) [ ] Major (Class 2) [ ] Minor (Class 3) [ ] Being Addressed (Class 4) [ ] No Action to be taken
Actual Cause	Methods Used to Verify Cause:
Initial Corrective Action Taken (including containment)	Methods Used to Verify Effects:

Final Action Taken	Area(s) Involved:	Functional Groups:
	[ ] Material	[ ] All
	[ ] Operator	[ ] Operations
	[ ] Equipment	[ ] Logistics
	[ ] Environmental	[ ] Technical
	[ ] Maintenance	[ ] Development
	[ ] Facilities	[ ] Marketing
		[ ] EH&S

Control Used to Prevent Reoccurrence	Disposition of NonConforming Material [ ] Use as is [ ] Return to Vendor [ ] Scrap [ ] Not Applicable

Responsibility: _____

QA/EHS Verification: _____ Date: _____

APPENDIX B

Corrective Action Cover Letter

TO: _____ DATE: _____ CA #: _____

The Management Review Board and/or Environmental Management Committee has assigned the responsibility for responding to the attached Corrective Action to you. The Risk Level for this corrective action is Class # _____ (see below). Please describe the corrective action you propose, the estimated date of its completion, and return this form to the Quality Assurance. This form is due back before the end of: _____.

Estimated Completion Date for Corrective Action: _____ and Proposed Corrective Action:

Class	QA	EH&S
1	A critical nonconformance or issue which shows a *major* deficiency in a system and which may have the potential for jeopardizing a standards certification. *To be completed within one week*	Any condition or practice with potential for: (a) permanent disability, loss of life or body part; (b) extensive loss of structure/equipment or material; and/or (c) extensive environmental/ecological damage. *To be completed within 1–12 hours*
2	For multiple nonconformances or issues within an area which may or may not have the potential for jeopardizing a standards certification. *To be completed within one month*	Any condition or practice with potential for: (a) serious injury/illness; (b) property damage that is destructive but less severe than Class 1; and/or (c) environmental/ecological damage less severe than Class 1. *To be completed within 12–36 hours*
3	Any nonconformance or issue which is considered a minor deficiency in a process and warrants a process improvement. *To be completed within three months*	Any condition or practice with potential for: (a) a non-disabling injury or illness; (b) non-disruptive property/equipment damage; and/or (c) nondestructive environmental/ecological damage. *To be completed within 36–72 hours*
4		Any other condition or practice, such as those involved in engineering modifications with a corrective action time to be discussed and agreed upon by the concerned parties.

- Corrective action request forms shall be made available throughout the company as a means to ensure open communication through all levels of the organization. Quality Assurance shall be responsible for maintaining collection boxes and collecting any requests on a day-to-day basis.
- All corrective action requests shalll be assigned a tracking number that is maintained on a database (Appendix C). The Quality Assurance Department will be responsible for maintaining the numbering system for all corrective actions, including environmentally related.
- The database will include the tracking number, the date the corrective action was issued, the responsible area, the responsible person, the nonconformance(s) noted, the assigned risk level(s), the estimated date(s) of completion and the actual completion date(s).
- The progress of open corrective actions shall be done by the Management Review Board in conjunction with the Environmental Management Committee.
- Metrics shall be established to track individual area performance against the required completion times (Appendix D). These metrics shall be reviewed on a periodic basis by the Management Review Board and the Environmental Management Committee. The metrics shall be posted for internal communication and objectives and targets shall be established on an annual basis to drive continual improvement of the system and improve auditing performance.
- Document Control shall be responsible for maintaining all records and files of completed corrective actions for external auditing purposes.

In Appendix D of the procedure, the first column in each section indicates the number of corrective actions assigned to a particular area for that particular risk class. The second column is the total number of days and the third column is the average days taken to complete all of that particular class. This allows management to assess which area is responding or not responding to corrective actions as required.

17.4.2 RISK ASSESSMENT

In Appendix D of the procedure, you also notice that I have combined both quality and environmentally related metrics. In reality, completion requirements due to differences in risk being evaluated, may require separate metrics for each. Response time for an environmental corrective action request needs to be watched much more carefully due to the potential liabilities involved. It is, therefore, important that the level of risk be evaluated much more closely before assigning a response classification. A Risk Assessment tool should be employed to determine what type of risk is involved, the potential impact and liability, and the probability that such a risk will happen. The formula involved is:

$$Risk = Severity \times Probability,$$

APPENDIX C
Corrective Action Database

CA #	Date	Assignee	Area	Source	Request	Class	Due Date	Est. Comp	Days to Comp.	Comp. Date
97-001	1/30/97	Joe S.	QC	ISO 9001 audit	Personnel are not certified to SOP #12345	3	4/30/97	3/30/97	43	3/15/97
97-002	2/15/97	Peter K.	Health & Safety	OSHA inspection	Machine guard does not adequately protect operator	1	2/15/97	2/15/97	1	2/15/97
97-003	3/18/97	Pat D.	Cell 1	ISO 14001 audit	Personnel are not aware of the significant environmental aspects	3	3/21/97	3/21/97	2	3/20/97
97-004	4/06/97	Jane E.	Cell 2	EHS inspection	Air emissions log is not being properly completed	1	4/06/97	4/06/97	4	4/10/97
97-005	5/15/97	Ellen R.	Warehouse	Mgmt. Inspection	SOP binders not available	3	6/15/97	5/31/97	15	5/30/97
97-006	6/28/97	Evelyn K.	CSC	Customer complaint	Shipment #Y2965B did not meet specifications	2	7/5/97	7/1/97	5	7/3/97
97-007	7/30/97	Tim M.	Doc. Control	ISO 9001 audit	History records for SOP #36927 were unable to be located	2	8/6/97	8/2/97	3	8/1/97
97-008	8/13/97	Mark S.	Technical	Operator	HV test unit continues to shock technicians	1	8/13/97	8/13/97	1	8/13/97
97-009	9/12/97	Joe Y.	Training	Internal audit	Training records do not include EMS communications training	3	12/12/97	11/15/97	99	12/20/97
97-010	9/22/97	Chuck C.	Environ.	Process Eng.	Consumption of IPA appears to be exceeding air permit limits	2	9/23/97	9/24/97	2	9/24/97

APPENDIX D
Corrective Action Metrics by Area

Area	Class 1	Completion Days		Class 2	Completion Days		Class 3	Completion Days		Class 4	Completion Days	
		Total	Avg. Days		Total	Avg. Days		Total	Avg. Days		Total	Avg. Days
QC							1	43	43.00			
Health & Safety	1	1	1.00									
Cell 1							1	2	2.00			
Cell 2	1	4	4.00									
Warehouse							1	15	15.00			
CSC				1	5	5.00						
Doc. Control				1	3	3.00						
Technical	1	1	1.00									
Training							1	99	99.00			
Environmental				2	2	2.00						
Totals	3	6	2.00	3	10	3.33	4	159	39.75	0	0	0.00

TABLE 17.2
Calculating Corrective Action Severity

| Risk Level | Area(s) Being Impacted or Consequences | | | | | Score |
	Legal	Injury	Process Loss	Business Interruption	Fines & Penalties	
Extreme						8–10
High						6–8
Serious						4–6
Moderate						3–4
Low						2–3
None						1–2

where risk is generally expressed as a percentage. Probability is generally calculated by determining the frequency of exposure (FE) to the risk multiplied by the likelihood of exposure (LE) to the risk. Thus, the formula is better expressed in the form:

$$Risk = Severity \times (FE \times LE)$$

Table 17.2 or some variation can be used to determine the severity score. In the boxes of Table 17.2, you can assess appropriate descriptions which will assist you or the assessor in determining an appropriate severity score. The intent is to decide which of the areas will be impacted and/or the consequences and then to note which one of them has the highest severity score. The highest severity score is then used in the risk assessment formula. In many cases, however, there may be more than one area that is impacted or at risk. An optional way to calculate the severity in such cases is to determine the score for all of the risks and then take an average value and then use this number as your severity score.

To calculate the probability you can use some variation of Table 17.3. Values for potential frequency of exposure are from 0 to 1 and likelihood of exposure are from 0 to 10. However, these values can be reversed and do not necessarily have to be as I have presented.

Once a risk score has been calculated, the next step is to equate this to the hazard classifications used in the corrective action process. Since there are four classifications, you can break the risk scores (1-100%) into four equivalent sections and equate them with the classes used on the corrective action form (Table 17.4). The breakdown of the risk score does not have to be in equal portions, but may be evaluated according to your own particular needs (i.e., an extremely high risk score may be from 60–100%).

17.5 PREVENTIVE ACTION

Both ISO 9001 and ISO 14001 require that your system include a process for utilizing preventive measures or a more proactive approach to identifying nonconformances instead of just waiting for them to happen and then responding to them (reactive). As mentioned earlier, ISO 14001 gives cursory mention of preventive

TABLE 17.3
Calculating Corrective Action Probability

Frequency of Exposure

Description	Score
Continuous throughout day	0.9–1.0
Every few hours	0.7–0.8
Every few days	0.6–0.7
Once per week	0.4–0.5
Once a month	0.2–0.3
Once a quarter	0.1–0.2
Every six months	0.05–0.1
Once a year	0.01–0.05
Decade or more	<0.01

Likelihood of Exposure

Description	Score
Extreme	9–10
High	8–9
Moderate	7–8
Low	6–7
Rare	5–6
Very rare	4–5
Extremely rare	3–4
None	1–3

TABLE 17.4
Comparison of Risk Score to Corrective Action Class

Risk Score	CA Class	EHS Response Time	QA Response Time
1–25	1	NA	NA
25–50	2	36–72 hours	3 months
50–75	3	12–36 hours	1 month
75–100	4	<12 hours	1 week

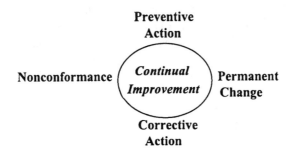

FIGURE 1. Continual improvement cycle.

action, but this does not mean it is not as important. In the environmental arena, a preventive action program will have a profound effect on your continual improvement process and your prevention of pollution efforts. In fact a continual improvement cycle can also look like Figure 1.

This figure shows that a preventive action is the starting point of the process that identifies any nonconformance. The nonconformance is placed into the corrective action process and out of this comes a permanent fix or change. The process leads back to the preventive action step.

The following list of activities can be used to identify nonconformances:

- Failure Modes and Effects Analysis (FMEA)
- Regulatory Compliance Audits
- Statistical Process Control (SPC)
- Internal Audits
- Design Reviews
- Employee Training and Certification (competence)
- Continuous Improvement Teams (CITs)
- Corrective Action Teams (CATs)
- Communication
- Emergency Preparedness Drills
- Monitoring
- Customer Complaints
- Service Reports

Although ISO 9001 stipulates a procedure for preventive action, it is not necessary to write such a procedure independent of the corrective action procedure. The requirements for preventive action can be integrated into the corrective action procedure in several locations (i.e., definition, responsibility, and procedure) and included in any process flow diagrams.

17.6   WHAT AUDITORS WILL LOOK FOR

For any corrective action process, it is important to ensure that a system is in place to handle nonconformances: the means to identify them (preventive actions), receive them, address them (corrective actions) and prevent them (permanent changes). An auditor will look for a system which has responsibility assigned and a system to ensure their timely completion per the requirements you establish.

18 Records

18.1 INTRODUCTION

When an auditing firm comes to appraise a business' operation for certification to either ISO 9001 or ISO 14001, the auditor will be requesting and evaluating physical evidence of conformance to the specified requirements. Management must make readily available records to demonstrate how effectively their organization is implementing the standards. These documents or "records" and the control of them is over and above the requirements of the document control process found under ISO 14001 (Element 4.4.5) and ISO 9001 (Element 4.5).

Both ISO 9001 and ISO 14001 have the potential for generating a large quantity of records, and without some form of control system, these records may become disorganized and/or unable to be readily located. As a result, the two standards have a requirement to write and implement a procedure to identify and maintain all quality and environmental records. These documents can be in either hardcopy (paper) or softcopy (computer/electronic) form as long as they are readily available and retrievable by personnel who need them.

18.2 COMPARISON OF THE STANDARDS

Table 18.1 shows a side-by-side comparison of the two standards and you will notice that they are almost identical. This table, however, has been expanded to include ISO 9004 and ISO 14004, the guideline documents for the two standards. By including them you will get a much broader picture of the requirements. In general, records management is fairly straightforward and, essentially, there is no reason for the two standards to be substantially different.

ISO 9001 and ISO 9004 contains some very specific requirements regarding customers and subcontractors which do not have a comparative requirement in ISO 14001. Specifically, ISO 9001 states that:

> pertinent quality records from the subcontractor shall be an element of these data and, where agreed contractually, quality records shall be made available for evaluation by the customer or the customer's representative for an agreed period.

ISO 9004 states that:

> policies should be established concerning availability and access of records to customers and subcontractors and pertinent subcontractor documentation should be included.

TABLE 18.1
Correlation of "Records" Requirements

ISO 9001		ISO 9004		ISO 14001		ISO 14004	
Section	Requirement	Section	Requirement	Section	Requirement	Section	Requirement
	The supplier shall establish and maintain documented procedures for …		The supplier shall establish and maintain documented procedures as a means for …		The organization shall establish and maintain procedures for …		The key features include …
4.16 (i)	identification	17.1	identification	4.5.3 (i)	identification	4.4.4	identification
4.16 (i)	maintenance and	17.1	maintenance	4.5.3 (i)	maintenance and	4.4.4	maintenance
4.16 (i)	disposition of quality records.	17.1	disposition of pertinent quality records	4.5.3 (i)	disposition of environmental records	4.4.4	disposition
4.16 (i)	collection, indexing, access, filing, storage	17.1	collection, indexing, filing, storage			4.4.4	collection, indexing, filing, storage
	Pertinent quality records from …				These records shall include …		
4.16 (ii)	the subcontractor shall be an element of these data.			4.5.3 (i)	training records and		
				4.5.3 (i)	the results of audits and		
				4.5.3 (i)	reviews		
	All quality records shall be …		All documentation should be …		Environmental records shall be …		
4.16 (iii)	legible	17.3	legible	4.5.3 (ii)	legible		
		17.3	readily identifiable	4.5.3 (ii)	identifiable and		
		17.3	dated, clean	4.5.3 (ii)	traceable		

				Environmental records shall be …			
4.16 (iii)	stored and			4.5.3 (ii)	stored and		
4.16 (iii)	retained	17.3	maintained	4.5.3 (ii)	maintained		
in such a way that they are …		quality records should be be …		in such a way that they are …			
4.16 (iii)	readily retrievable in facilities that provide suitable environment to	17.1 17.2 17.3	retrieval / readily retrievable for analysis / retrievable	4.5.3 (ii)	readily retrievable and	4.4.4	retrieval
4.16 (iii)	prevent damage or deterioration and to prevent loss.	17.2	protected in suitable facilities from damage, loss and deterioration	4.5.3 (ii)	protected against damage deterioration and loss		
Retention times of quality records shall be …			Quality records should be retained for a specified time …	Their retention times shall be …		4.4.4	retention
4.16 (iii)	established and	17.3	defining retention times	4.5.3 (ii)	established and		
4.16 (iii)	recorded.			4.5.3 (ii)	recorded.		
Quality records shall be …		The quality system should require that…		Records shall be …			
4.16 (ii)	maintained to demonstrate conformance to specified requirements and the effective operation of the quality system.	17.2	sufficient records be maintained to demonstrate conformance to specified requirements and verify effective operation of the quality system.	4.5.3 (iii)	maintained, as appropriate to the system and to the organization, to demonstrate conformance to the requirements of this International Standard.		

18.3   WHAT IS A RECORD?

A record is defined in the dictionary as a "collection of related items of information (as in a database) treated as a unit." Although a *document* control program and *records* control program appear to have some overlap, there are basic differences between them. *Document control* is mainly concerned with the establishment of procedures that ensure the quality process will function in accordance with the International Standards or any other codes and regulatory requirements. It also determines how procedures are to be created and modified or changed. In contrast, *records control* has its primary focus on the supporting documentation that verifies whether or not the procedures are effective. Records provide information to monitor and measure the quality or environmental management system and to implement the corrective action process.

18.3.1   TYPES OF RECORDS

The following are examples of quality and environmental records which require control:

- Legislative and regulatory requirements
- Inspection reports
- Testing data
- Permits to operate
- Qualification reports
- Environmental aspects and associated impacts
- Training reports
- Internal audit reports
- External audit reports
- Monitoring data
- Maintenance reports
- Calibration data
- Inspection reports
- Cost reports
- Customer inquiries and questionnaires
- Customer complaints
- Details of nonconformances
- Material Review reports
- Drawings
- Material/product specifications
- Management reviews
- Meeting minutes
- Supplier and subcontractor reviews and information and
- Procedures and manuals

Records must include any charts and databases (spreadsheets) that assist in monitoring the system. As you can see this is a very long list and is by no means exhaustive.

TABLE 18.2
Records Retention Requirements

Record Type	Retention Time (yrs.)
Environmental: hazardous waste tracking logs	3
Environmental: Material Safety Data Sheets (MSDS)	30
Environmental: test results, waste analysis, etc.	3
Environmental: air emissions monitoring	2
Environmental: aspects, impacts and programs	5
Management Reviews: meeting minutes, objectives progress, etc.	3
Contracts: agreements with customers (i.e., POs, contracts, etc.)	7 after warrant expiration
Contracts: customer order correspondence	4
Customer: inquiries, questionnaires and complaints	7
Design Reviews: reports	permanent
Design Reviews: laboratory notebooks	permanent
Design Reviews: data, drawings, specifications	15 after discontinued
Manufacturing: process data (i.e., charts, etc.)	3
Training: education and certification records	10 after termination
Training: job descriptions and responsibilities	5 after superseded
Purchasing: Supplier purchase orders	7
Purchasing: Supplier evaluations and quality data	15
Purchasing: Supplier corrective actions	7
Audits: external and internal	7
Calibration and Maintenance	15

18.4   RETENTION REQUIREMENTS

In most cases, the quality system will not be affected by specific regulatory requirements for retaining records. Contractual arrangements with suppliers, subcontractors
and customers will generally stipulate those requirements. Table 18.2 gives examples
of suggested retention times for various quality documents. Included in this table are
retention times for various environmental records and most of them are timeframes
established by law. All other records, such as meeting minutes, are generally at the
discretion of an organization's management and I have included arbitrary retention times
for this example. This can be included in the procedure in the next section.

18.5   IDENTIFICATION, COLLECTION, INDEXING, STORAGE, ACCESS, FILING, MAINTENANCE, AND DISPOSITION

In order to describe all of these requirements, I will again write a procedure based
on the format utilized throughout this book. Again, this is just an example, and you
must tailor yours specifically to the needs of your own organization.

Purpose   This procedure defines the control requirements for records within the organization and outlines the methods used in the identifying, collecting, filing, storage, maintenance, retrieval and disposition of quality and environmental records.

Scope   This procedure applies throughout the company and all personnel who generate and maintain records of any type (see list in Appendix B).

Definition of terms

- A *Record* is any unit of information that is recorded either on paper, microfilm, microfiche, computer tape or disk or any other media.
- An *Environmental Record* is a record maintained for either legal compliance (EPA), international standards or company requirements
- A Quality Record is a record that is evidence of conformance to product specifications, technical requirements, contracts and regulatory authorities (ISO 9001).

Referenced documents

- ISO 14001, Element 4.5.3, Records
- ISO 9001, Element 4.16, Control of Quality Records

EH&S precautions   NA

Precedence   If there is a conflict between this procedure and a customer requirement or regulatory agency, the latter shall take precedence.

Responsibilities

- Each department or area which maintains records of any type shall be responsible for complying with the requirements of this procedure.
- Each department or area is responsible for proper labeling and packaging of records and contacting storage vendor for permanent storage.

Procedure

Collection and Filing

- All records are collected at the point of origin and filed. Types of records are listed in Appendix A.

Retention

- Retention times for records are listed in Appendix A.
- All records are to be retained for the minimal time period indicated.
- A master file shall be kept by each department or areas as to what records have been sent to permanent storage.

Transfer to Permanent Storage

- Fill out proper storage vendor forms.
- Fill out proper labels and affix to each package, box or carton.

Retrieval from Permanent Storage

- Fill out proper retrieval form from storage vendor.
- Contact storage vendor for transfer.

Disposal of Records

- Complete proper forms provided by the storage vendor.
- Vendor will dispose of records under proprietary procedures.

18.6 WHAT AUDITORS WILL LOOK FOR

There are four key questions which an auditor will consider when appraising compliance to this section. First of all, is there evidence that an organization has given consideration to the types of records which must be controlled? Second, has the organization developed a documented procedure for managing records? Third, has the organization identified and tracked key performance indicators (along with corresponding data) to monitor and measure progress against objectives and targets and, ultimately, compliance with the quality and/or environmental policy? Fourth, has the organization developed a system to make the records and other information available to all personnel? If you can safely answer "Yes" to all of these, the audit results should not result in a nonconformance.

19 Environmental Management Systems Audit

19.1 INTRODUCTION

An audit is a periodic and predetermined evaluation of an organization's compliance to a set of standards or legal requirements. With standards, it is generally a very detailed appraisal of how the organization is meeting their policy commitments through the evaluation of the procedures and processes that drive the overall system.

The auditing process is considered by many to be a very important process and is the organization's best tool to identify nonconformances internally. Through the internal audit process, an organization can very effectively reduce or eliminate the potential for nonconformances that may be found by an external auditing body (i.e., ISO registered auditor or a regulatory agency). All of the elements of the ISO 9001 and ISO 14001 standards should be audited on a periodic basis with special attention being given to those activities which have actual or the greatest potential to impact your operation — whether quality or environmentally related. ISO 9004, in Section 5.4.2, also states that *"other factors necessitating audits can be organizational changes, market feedback, nonconformity reports and surveys.."* In other words, do not focus the need to do audits on just trying to meet ISO 9001 or ISO 14001 requirements.

19.2 COMPARISON OF THE STANDARDS

As I have done in the previous chapter, I would like to include the information provided in ISO 9004 in comparing the two standards. ISO 9004 contains valuable information and gives much greater detail on what an auditing system should include. It has been my experience that the guideline document (ISO 9004) has been much more valuable in ascertaining, establishing, and implementing the requirements of a Quality Management System. Section 5.4 of ISO 9004 can provide an excellent framework for building both your ISO 9001 and ISO 14001 auditing system starting with the development of the procedure. Structured on the same format as Table 18.1, Table 19.1 provides a side-by-side comparison of all three documents, but does not include ISO 14004, the guideline document for ISO 14001. Additionally, the table does not include some ISO 9004 information pertaining to the extent of audits (5.4.3) and specific audit reporting requirements (5.4.4).

TABLE 19.1
Correlation of 'Auditing' Requirements

ISO 9001, Element 4.17		ISO 9004		ISO 14001, Element 4.5.4	
Section	Requirement	Section	Requirement	Section	Requirement
The supplier shall establish and maintain ...		The audit program should cover ...		The organization shall establish and maintain ...	
(i)	documented procedures for planning and implementing internal quality audits	5.4.2 (c)	documented procedures for carrying out audits	(i)	(a) program(s) and procedures for periodic environmental audits to be carried out,
to ...				in order to ...	
	verify whether quality activities and related results ..		Audits should be planned and carried out to determine if the activities and related results of the organization's quality system ...	(a)	determine whether or not the environmental management system ...
4.17 (i)	comply with planned arrangements and	5.4.1	comply with planned arrangements and	(a)(1)	conforms to planned arrangements for environmental management including the requirements of this International Standard; and
4.17 (i)	to determine the effectiveness of the quality system.	5.4.1	to determine the effectiveness of the quality system.	(a)(2)	has been properly implemented and maintained.
4.17 (iii)	recorded and brought to the attention of personnel having responsibility in the area audited.	5.4.2 (c)	recording and reporting the results of the quality audit	(b)	provide information on the results of audits to management.
Internal quality audits shall be scheduled on ...		All elements should be internally audited and evaluated on a regular basis, considering		The organization's audit program, including any schedule, shall be ...	
4.17 (ii)	the basis of the status and importance of the activity to be audited and	5.4.1	the status and importance of the activity to be audited.	(ii)	based on the environmental importance of the activity concerned and

	5.4.5	Implementation and effectiveness of corrective actions resulting from previous audits should be assessed and documented.	(ii)	the results of previous audits.
		The audit program should cover …		In order to be comprehensive, the audit procedures shall …
	5.4.2 (a)	planning and	(ii)	cover the audit scope
	5.4.2 (a)	scheduling the specific activities and areas to be audited.	(ii)	frequency
			(ii)	methodologies
4.17 (ii)	5.4.2 (b)	assignment of personnel with appropriate qualifications to conduct audits reaching agreement on timely corrective actions on the deficiencies found during the audit	(ii)	as well as the responsibilities and requirements for conducting audits and reporting results.
	5.4.2 (c)	shall be carried out by personnel independent of those having direct responsibility for the activity being audited.		
4.17 (iii)		The management personnel responsible for the area shall take timely corrective action on deficiencies found.		

19.3   A PREASSESSMENT EVALUATION

One type of audit that can be done either internally or by an external organization is a PreAssessment Evaluation or a "gap" analysis. This is an "audit" of your operation performed prior to implementing a formal standards program. A preassessment of your process against either ISO 9001 or ISO 14001 can assist management in understanding exactly where the deficiencies are in meeting the standard's requirements. It can also assist management in determining when a preliminary or Stage 1 audit can be done by the auditing body.

In Appendix F you will find an ISO 14001 Self-Assessment checklist that can be used either internally or by an external group. This checklist breaks the ISO 14001 Environmental Management Standard into its basic components and asks if there is any evidence of compliance to the requirement. If a procedure is specifically required, it can be noted and other specific questions are asked to assist in verifying the section questions. This checklist is just one of many preassessment tools available, but care must be taken when evaluating any preassessment tool. The checklist in Appendix F is the ISO 14001 Standard itself and is, therefore, the best type of checklist you should consider — when in doubt, use the standard itself!

19.4   INTERNAL AUDITS

The bulk of this chapter will be spent on the internal audit process because of the extensive program that must be developed to comply with the requirements. Internal audits can be of two types: area-specific and system-wide.

19.4.1   AREA-SPECIFIC INTERNAL AUDITS

This type of audit has its primary focus on a specific area to appraise how that area is complying with all or a major part of the ISO 9001 and/or ISO 14001 Standards.

Below is an example of an internal auditing checklist that can be used to appraise an area's compliance to both ISO 9001 and ISO 14001. The checklist contains the requirement for each element and provides an auditor information concerning what should be looked at and the specific items or activities associated with it. The checklist uses the ISO 9001 elements as the foundation and integrates the various ISO 14001 elements which have similar bearing on the subject. Most of the questions can either be applied specifically to quality concerns or to environmental issues. Some of the questions are quality or environmental specific. The ISO 9001 elements in this checklist are not necessarily in order to conserve space.

ISO 14001/9001 INTERNAL AUDIT CHECKLIST

ISO 9001 Element 4.1 Management Responsibility
ISO 14001 Element 4.2 Environmental Policy
ISO 14001 Element 4.4.1 Structure and Responsibility
ISO 14001 Element 4.4.3 Communication

Requirement   The environmental and/or quality policy shall be defined, understood, implemented and maintained.

Evaluate   Quality and Environmental Policy(s)

Ask sampling of operators about the quality and/or environmental policy. Is it understood?	Comments:
How does it relate to what they are doing?	Comments:

Evaluate   Environmental Policy

Is there a commitment to *Continual Improvement,*	
Environmental Legislation and Prevention of Pollution and/or any other requirements?	Comments:

Evaluate   Structure and Responsibility

Have roles, responsibility and authorities been defined and documented?	Comments:
Is there evidence that adequate resources have been assigned to implement the system?	Comments:
Are personnel aware of their responsibilities in terms of product quality and environmental management?	Comments:
Is there evidence of communication between all levels of the area/organization?	Comments:

Evaluate   Customer Service procedure for handling inquiries

Is (are) the policy(s) available to external interested parties (i.e., public, customers, etc.)?	Comments:

ISO 9001 ELEMENT 4.12 INSPECTION AND TEST STATUS

Requirement   Status of inspections and tests shall be maintained for items as they progress through various processing steps and records show who released conforming product.

Evaluate   Status of lots in production

Is the current status of lots in production known and the record updated and available?	Comments:

Evaluate   Process record

Is there clear evidence that all process steps and tests have been done (e.g., operator signature)?	Comments:

ISO 9001 Element 4.2 Quality System
ISO 14001 Element 4.4.4 EMS Documentation
ISO 14001 Element 4.4.6 Operational Control
ISO 14001 Element 4.3.4 Environmental Management Programs

Requirement   Documented procedures shall be prepared and implemented.

Evaluate   Implementation

Is there evidence that all processes are being done as specified in the documents?	Comments:

Evaluate   Documented Procedures

Are procedures available for each process step, including maintenance?	Comments:
Are consequences identified in procedures for deviations to established requirements?	Comments:

Evaluate   Procedures for goods and services

Are they communicated to suppliers and contractors?	Comments:

Requirement   A Quality and/or Environmental Manual shall be prepared.

Evaluate   Manual(s)

Does it (they) describe how the quality and/or environmental requirements shall be met and planned?	Comments:
Does it (they) provide direction to related documentation?	Comments:

Requirement   Programs shall be established for achieving quality and environmental objectives and targets.

Evaluate   Program Documentation

Is there evidence that responsibility has been assigned?	Comments:
Has the means and time-frame to achieve the objective and target been identified?	Comments:

ISO 9001 ELEMENT 4.3 CONTRACT REVIEW
ISO 14001 ELEMENT 4.3.1 ENVIRONMENTAL ASPECTS
ISO 14001 ELEMENT 4.4.6 OPERATIONAL CONTROL

Requirement   Customer requirements must be adequately defined and documented.

Evaluate   Contracts

Are contracts reviewed in accordance with established procedures?	Comments:
Is there a timely review (measured in days vs. weeks or months) of contracts?	Comments:

Evaluate   Purchase Orders (customer)

Are POs adequately defined, reviewed, and all issues resolved?	Comments:
How are customer specifications and engineering standards controlled?	Comments:
Are they dated with the receipt date?	Comments:

Evaluate   Implementation

How are customer requirements communicated to the appropriate manufacturing personnel?	Comments:
Are customers' special characteristics communicated to appropriate personnel for incorporation into manufacturing documents?	Comments:

Evaluate   Procedure for identifying environmental aspects

Are contractors considered to be actual or potential impacts?	Comments:

ISO 9001 Element 4.4 Design Control
ISO 14001 Element 4.3.1 Environmental Aspects
ISO 14001 Element 4.3.2 Legal and Other Requirements

Requirement   Design output, including crucial product characteristics, shall be documented.

Evaluate   Technical Group activities

Are product development activities (projects) planned and responsibilities assigned?	Comments:
Is there evidence that the plans are being revised as needed (development plans regularly updated)?	Comments:
Are feasibility reviews conducted to confirm compatibility of design with the manufacturing process, including capacity planning and utilization?	Comments:
Will the new product or process create a new environmental impact?	Comments:

Evaluate   Procedure for product development

Are all requirements, particularly design inputs, design outputs, crucial product characteristics, and design verification included in the design reviews?	Comments:
Do the requirements for the above include evaluating the environmental aspects of the product development?	Comments:
Will a new product or process impact existing environmental objectives and targets?	Comments:

ISO 9001 Element 4.8 Product Identification and Traceability
ISO 14001 Element 4.3.2 Legal and Other Requirements

Requirement   The product shall be identified and traceable by item, batch, or lot number during all stages of production, delivery, and installation.

Evaluate   Check Product or Material Identification and Traceability during all internal audits.

Comments:

Evaluate: Final and In-Process Labels or Identification

Are the documented instructions being followed for labeling product during production?	Comments:
Is product identified at all stages of production?	Comments:
Are the requirements of final labels identified, including environmental?	Comments:

Evaluate   Equipment/Tools

Are all pieces of equipment/tools identified and labeled with a tracking number?	Comments:

ISO 9001 Element 4.5 Document Control
ISO 14001 Element 4.4.5 Document Control

Requirement   Generation, distribution, and changes to documents must be controlled.

Evaluate   Control Books

Check revision levels of a sampling of documents for latest revisions in Document Control?	Comments:
How are changes to documents communicated to the group? (i.e., Read and Understood form, training roster, etc.?)	Comments:

Evaluate   Uncontrolled or Posted Documents

Are uncontrolled or reference copies being used to control processes?	Comments:

Evaluate   Machine & Tooling Drawing

Are they available, controlled, and the correct revision?	Comments:

Evaluate   Sample of other documents such as Training Records, Inspection Reports, etc.

Are they included in the document control procedure and how are they specifically being controlled?	Comments:

ISO 9001 Element 4.16 Quality Records
ISO 14001 Element 4.5.3 Records

Requirement   Quality records shall be identified, collected, indexed, filed, stored, maintained, and dispositioned.

Evaluate   All documents/forms/logbooks/reports on which product/process results have been written.

Do records exist for identification, collection, indexing, filing, storing, maintaining, and dispositioning?	Comments:
Where are Production Packets stored?	Comments:
Is there a clear method of retrieval?	Comments:
Are quality and environmental records stored in accordance with established procedures?	Comments:
Do the procedures identify all types of records covered by this requirement?	Comments:

ISO 9001 ELEMENT 4.6 PURCHASING
ISO 14001 ELEMENT 4.3.1 ENVIRONMENTAL ASPECTS
ISO 14001 ELEMENT 4.4.6 OPERATIONAL CONTROL

Requirement Suppliers and contractors shall be evaluated for their ability to meet specified quality and environmental requirements.

Evaluate Ordering Information (Requisitions)

Are all requirements given to Purchasing clearly defined both quality and environmental?	Comments

Evaluate Ordered Materials

Is there evidence that materials were evaluated for their ability to provide to requirements (1st article inspection) and to minimize any potential environmental impact?	Comments:
Are there any environmental legislation restrictions and are they noted and understood?	Comments:

Evaluate Supplier List

Is there a current list of approved suppliers?	Comments:

Evaluate Supplier's Quality and Environmental System

Have the supplier's quality and environmental systems been assessed and are there records of demonstrated capabilities?	Comments:
Is the supplier's delivery performance evaluated?	Comments:
Is excess or premium freight monitored?	Comments:

Evaluate Purchase Orders

Do they contain enough detail and contain latest revision drawings?	Comments:
Do they contain requirements to include Material Safety Data Sheets with all chemical and raw material shipments?	Comments:

Evaluate Corrective Actions

Have corrective actions been taken and have they been effective?	Comments:

ISO 9001 ELEMENT 4.9 PROCESS CONTROL
ISO 14001 ELEMENT 4.3.1 ENVIRONMENTAL ASPECTS
ISO 14001 ELEMENT 4.4.6 OPERATIONAL CONTROL
ISO 14001 ELEMENT 4.5.1 MONITORING AND MEASUREMENT

Requirement Process shall be carried out under documented instructions.

Evaluate Process Control Plan, FMEA, and Process Flow Charts

Are Process Control plans updated, void of specifications and reflect current practices? Are FMEA's and Process Flow Charts current?	Comments:
Do FMEA's and Process Control Plans contain special characteristics that identify minor, major or critical measurements (quality and environmental)?	Comments:

Evaluate Production Planning Procedures

Are production planning procedures being followed?	Comments:

Evaluate Manufacturing Procedures

Is there evidence that operators understand and are following procedures to control product quality and minimize the processes impact on the environment?	Comments:
Are there procedures for aspects of the operation which have a significant impact on the environment?	Comments:

Evaluate Criteria Boards

Are criteria boards available and current?	Comments:

Evaluate Software

Is software controlled according to the requirements of QAP 22336?	Comments:

Evaluate Use of Computers

Are there written procedures to cover the use of computers to track product or process information?	Comments:

Evaluate Records

Do all processes that cannot be verified after the fact have records that indicate appropriate process control?	Comments:

ISO 9001 ELEMENT 4.10 INSPECTION AND TESTING
ISO 14001 ELEMENT 4.3.1 ENVIRONMENTAL ASPECTS
ISO 14001 ELEMENT 4.5.1 MONITORING AND MEASUREMENT

Requirement   Incoming materials shall be inspected or verified before use and in-process testing shall be performed according to procedures, and final inspection and testing shall be performed prior to release of finished product.

Evaluate   Incoming Production Material

Has all incoming productive material been verified prior to use?	Comments:
Where certifications are supplied, are test results provided?	Comments:
On incoming inspection, do suppliers submit statistical data as required?	Comments:
Has the regulatory status of all incoming raw materials and chemicals been established and identified?	Comments:

Evaluate   In-process testing

Is in-process testing being done to documented instructions including statistical-based sampling and appropriate pass/fail criteria?	Comments:

Evaluate   Final inspections

Are all final inspections completed prior to release of the product, unless otherwise documented?	Comments:

Evaluate   Production folder

Are production folders (run card, router, MPF) complete?	Comments:

Evaluate   Log, notebooks, data sheets, or test records

Are all product records complete?	Comments:
Are logbooks maintained as indicated by the procedures?	Comments:

Evaluate   Reviewed documents

If required, does evidence exist (e.g., stamp, initials, etc.) that review has been done and by authorized personnel?	Comments:

ISO 9001 ELEMENT 4.11 INSPECTION, MEASURING, AND TEST EQUIPMENT
ISO 14001 ELEMENT 4.5.1 MONITORING AND MEASUREMENT

Requirement   Equipment shall be controlled, calibrated, and maintained.

Evaluate   Equipment/gages/jigs/fixtures

Is there a valid calibration sticker, or a "not to be calibrated" label attached if equipment is not being used to control the process?	Comments:

Evaluate   Equipment affecting product quality or an environmental aspect.

Is all equipment affecting product quality and the environment included on a calibration schedule and a maintenance schedule?	Comments:

Evaluate   Calibration records

Do calibration records show evidence of review for calibration validity?	Comments:

Evaluate   Calibrated equipment

Is calibrated equipment in a suitable environment and show no evidence of tampering?	Comments:
Is equipment used by an outside vendor calibrated NIST, ANSI, or ISO requirements?	Comments:

Evaluate   Measurement validity

What is the procedure for product or process review if equipment is found out of calibration?	Comments:

Evaluate   Evidence of equipment capability?

Look for: Is there evidence that equipment is capable of required accuracy?	Comments:

Evaluate   In-House calibration

Where calibration is done in the division, are there calibration procedures, a log indicating calibration was done, and records of test results?	Comments:

Evaluate   Maintenance

Is there a procedure for scheduled maintenance for all equipment and tooling for calibrations, evaluation, or maintenance?	Comments:
What records are available to demonstrate preventive maintenance schedules?	Comments:
Are maintenance records evaluated to reduce downtime?	Comments:
Are spare parts inventory kept on key equipment?	Comments:

Evaluate   Test hardware or software

Is test hardware or software being used, checked for accuracy?	Comments:

ISO 9001 Element 4.13 Control of Nonconforming Product
ISO 14001 Element 4.3.1 Environmental Aspects
ISO 14001 Element 4.3.2 Legal and Other Requirements
ISO 14001 Element 4.5.2 Nonconformance/Corrective
and Preventive Action

Requirement   Review and disposition of nonconforming product shall be accomplished in a formal manner. Nonconforming product shall be controlled to prevent inadvertent use or installation.

Evaluate   Product Discrepancy Reports

Are reports filled out completely?	Comments:
Is proper environmental paperwork with the product or has it been identified previously (i.e., TSCA status)?	Comments:

Evaluate   Nonconforming Product

Is nonconforming material labeled correctly and segregated in a holding area?	Comments:

Evaluate   Records for disposition of material

Has dispositioning of material been done with the required approvals and is the approval dated?	Comments:
Are the nonconformance reports used to quantify and analyze non-conforming product and to establish a prioritized reduction plan?	Comments:

Evaluate   Product nonconformance logbooks

Are logbooks maintained?	Comments:
How long are the reports kept in the book? What is done after that time?	Comments:

Evaluate   Nonconformance forms

Is the latest revision of the form being used?	Comments:

ISO 9001 ELEMENT 4.14 CORRECTIVE AND PREVENTIVE ACTION
ISO 14001 ELEMENT 4.5.2 NONCONFORMANCE/CORRECTIVE AND PREVENTIVE ACTION

Requirement Problem causes should be identified, corrected, and the effectiveness of the correction assessed.

Evaluate Corrective Actions

Are corrective actions being written on the nonconformance reports?	Comments:

Evaluate Management meeting minutes

Is management maintaining evidence (meeting minutes) that they are reviewing nonconformance reports?	Comments:

Evaluate Department teams tracking nonconformance reports

Is there evidence that areas/departments are using the corrective actions from nonconformance reports to prevent future recurrence?	Comments:

Evaluate Corrective actions from previous audits

Is there evidence that corrective actions from previous audits have been implemented in a timely manner?	Comments:
Is there evidence that implemented corrective actions from the previous audit(s) have been effective?	Comments:

Evaluate Preventive Actions

Is there evidence that preventive action measures are being utilized?	Comments:

ISO 9001 ELEMENT: 4.17 INTERNAL QUALITY AUDITS
ISO 14001 ELEMENT 4.5.4 ENVIRONMENTAL MANAGEMENT AUDIT

Requirement Audits shall be planned and performed, results communicated to management, deficiencies corrected.

Evaluate Are self-audits being performed via the CITs?

Is there evidence of proactive continual improvement?	Comments:
Is there evidence of department-generated 'Defects and Barrier' Logs?	Comments:
Does the area maintain a safe and clean working environment?	Comments:
Are audit results reported to management?	Comments:
Does the audit process conform to planned arrangements under ISO?	Comments:

ISO 9001 ELEMENT 4.15 HANDLING, STORAGE, PACKAGING, PRESERVATION AND DELIVERY
ISO 14001 ELEMENT 4.3.1 ENVIRONMENTAL ASPECTS
ISO 14001 ELEMENT 4.3.2 LEGAL AND OTHER REQUIREMENTS

Requirement Handling, storage, packaging, preservation, and marking processes shall be controlled.

Evaluate Written instructions

Are there written instructions for handling and packaging where applicable?	Comments:
Do these written instructions take into consideration any special environmental requirements?	Comments

Evaluate Final Product labels

Are there written procedures for labeling final product, including eco-labeling requirements?	Comments:

Evaluate Maintaining product quality

Are there controls for maintaining the quality of product after final inspection?	Comments:

Evaluate Checks for product deterioration

Are there specific checks for product deterioration including shelf life, physical damage, water damage, etc. for product in stock?	Comments:

ISO 9001 ELEMENT: 4.20 STATISTICAL TECHNIQUES
ISO 14001 ELEMENT 4.5.1 MONITORING AND MEASUREMENT

Requirement   Establish procedures (where appropriate) for identifying adequate statistical techniques required for verifying the acceptability of process capability, product characteristics and environmental performance.

Evaluate   Statistical Techniques

Has the statistical techniques method been identified?	Comments:
Do all inspections have a statistical basis and follow the sample plan requirements in the procedures?	Comments:
Where specifications are being specified, are they being used correctly?	Comments:
Do the statistical techniques determine the acceptability of the process or product?	Comments:
Do the environmental statistics verify the requirement to prevent pollution?	Comments:
Are significant aspects of the operation being monitored and measured?	

ISO 9001 ELEMENT 4.18 TRAINING
ISO 14001 ELEMENT 4.4.2 TRAINING, AWARENESS AND COMPETENCE

Requirement   Training needs shall be identified, training provided to ensure competence and records maintained.

Evaluate   Training records

Are training needs identified and the training done?	Comments:
Is the training effectiveness periodically evaluated?	Comments:

Evaluate   Personnel certification and competence

Are all personnel (other than those in training) certified to the processes they are assigned?	Comments:
Are personnel aware of the consequences of not following established procedures?	Comments:

Evaluate   Training records maintenance

Are Personnel certification records maintained in accordance with established procedures?	Comments:
Is the training level of each employee known for a particular operation?	Comments:

Evaluate   Emergency Preparedness and Response

Are personnel aware of their responsibilities during an emergency?	Comments:
Are employees aware of their responsibilities when an environmental emergency occurs?	Comments:

19.4.2 SYSTEM INTERNAL AUDITS

This type of audit has its focus on specific sections or elements of a standard. In other words, an auditor will be assigned to conduct an appraisal of how an entire organization is doing in complying with one specific requirement of the standard. An example would be when an auditor focuses only on the document control aspect of an organization to determine if all areas understand and are implementing the document control requirements as planned. The checklist provided above can also be used for this type of audit

19.4.3 AUDIT SUMMARY REPORT

Once the audit has been conducted, the next step is to write a summary report which details the overall findings (both accomplishments and nonconformances). The report should include a condensation of the findings and describe the scope of the audit. The following is an example of a summary report:

EXAMPLE OF AN ISO AUDIT SUMMARY REPORT

QUALITY ASSURANCE AND ENVIRONMENTAL AUDIT SUMMARY REPORT

Audit Scope:
(indicate whether this audit covers all ISO elements within a defined area or is an evaluation of a specific ISO element over the entire organization)

Audit date:

Report date:

Escort Team:
(indicate which personnel accompanied the audit team and/or personnel who were interviewed to verify the questions)

Audit team:

Contents: Introduction
              Summary of Nonconformances
              Audit Results

INTRODUCTION

This report documents the audit performed on _____ in department or area _____.   (date)

The purpose of the audit was to evaluate current practices against requirements described by procedures and the organization's Quality and Environmental Manuals. The Quality and Environmental Manuals include corporate (divisional, group, etc.) policies, as well as requirements from ISO 9001:1009(E) and ISO 14001:1996(E). When there was a discrepancy between the audit finding and a written requirement, a nonconformance was noted.

The audit report is structured to provide feedback on the effectiveness of processes in complying with ISO 9001 and ISO 14001. Since the audit evaluated only parts of certain quality system and environmental system elements, lack of a nonconformance does not necessarily signify total compliance to an element of ISO 9001 and/or ISO 14001.

The audit team appreciates the support and openness extended by the escort(s) during the audit. Because this is an internal evaluation, the audit report does not mention all the items that were found to be in conformance with ISO 9001 and/or ISO 14001. At the risk of giving a distorted perspective of the systems correctly implemented in the area, the report is focused on those items that are noncompliant. It is provided by Quality Assurance and Environmental Management, as a means of sustaining the Organization's ISO 9001 and ISO 14001 Certifications, while providing support and direction on how to improve and strengthen our business practices in accordance with ISO 9001 and ISO 14001.

If there are any questions or comments, please contact the appropriate member of the audit team. Otherwise, it is anticipated that the area will formulate a plan that addresses all the nonconformances noted, and submit the plan to Quality Assurance and/or Environmental Management within 14 days from the receipt of the report/corrective action form(s).

The quality and environmental elements from ISO 9001 and ISO 14001 that were reviewed are as follows:

Summary of Areas Covered

Element	ISO 9001-Based Elements	Audited Yes	No
4.1	Quality Policy		
4.2	Quality System		
4.3	Contract Review		
4.4	Design Control		
4.5	Document and Data Control		
4.6	Purchasing		
4.8	Product Identification and Traceability		
4.9	Process Control		
4.10	Inspection and Testing		
4.11	Control of Inspection, Measuring, and Test Equipment		
4.12	Inspection and Test Status		
4.13	Control of Nonconforming Product		
4.14	Corrective and Preventive Action		
4.15	Handling, storage, packaging, preservation and delivery		
4.16	Control of Quality Records		
4.17	Internal Quality Audits		
4.18	Training		
4.20	Statistical Techniques		
	ISO 14001-Based Elements	Yes	No
4.2	Environmental Policy		
4.3.1	Environmental Aspects		
4.3.2	Legal and Other Requirements		
4.3.3	Objectives and Targets		
4.3.4	Environmental Management Programs		
4.4.1	Structure and Responsibility		
4.4.2	Training, Awareness and Competence		
4.4.3	Communication		
4.4.4	Environmental Management System Documentation		
4.4.5	Document Control		
4.4.6	Operational Control		
4.4.7	Emergency Preparedness and Response		
4.5.1	Monitoring and Measurement		
4.5.2	Nonconformance and Corrective & Preventive Action		
4.5.3	Records		
4.5.4	Environmental Management System Audit		
4.6	Management Review		

Summary of Nonconformances

Element	ISO 9001-Based Elements	Nonconformances Found? Yes	No
4.1	Quality Policy		
4.2	Quality System		
4.3	Contract Review		
4.4	Design Control		
4.5	Document and Data Control		
4.6	Purchasing		
4.8	Product Identification and Traceability		
4.9	Process Control		
4.10	Inspection and Testing		
4.11	Control of Inspection, Measuring, and Test Equipment		
4.12	Inspection and Test Status		
4.13	Control of Nonconforming Product		
4.14	Corrective and Preventive Action		
4.15	Handling, storage, packaging, preservation and delivery		
4.16	Control of Quality Records		
4.17	Internal Quality Audits		
4.18	Training		
4.20	Statistical Techniques		
	ISO 14001-Based Elements	Yes	No
4.2	Environmental Policy		
4.3.1	Environmental Aspects		
4.3.2	Legal and Other Requirements		
4.3.3	Objectives and Targets		
4.3.4	Environmental Management Programs		
4.4.1	Structure and Responsibility		
4.4.2	Training, Awareness and Competence		
4.4.3	Communication		
4.4.4	Environmental Management System Documentation		
4.4.5	Document Control		
4.4.6	Operational Control		
4.4.7	Emergency Preparedness and Response		
4.5.1	Monitoring and Measurement		
4.5.2	Nonconformance and Corrective & Preventive Action		
4.5.3	Records		
4.5.4	Environmental Management System Audit		
4.6	Management Review		

19.5   INTERNAL AUDIT PROCEDURE

Both ISO 9001 and ISO 14001 require a written, documented procedure which details the requirements for internal auditing. The procedure should include the scope, frequency, methodologies, responsibilities and requirements for conducting audits and reporting results. Below is an example of an integrated procedure that can be used to ensure all of these requirements are met.

OPERATING PROCEDURE FOR INTERNAL AUDITS

Purpose   This procedure prescribes the methodology to be used in conducting internal quality and environmental audits as a means to determine whether or not the management system conforms to planned arrangements and has been implemented and maintained according to the ISO 9001 and ISO 14001 Standards.

Scope   This procedure defines how internal audits are to be conducted and how nonconformances and subsequent corrective actions are managed.

Definition of terms

- **ISO 14001** is the worldwide environmental management system (EMS) standard.
- **ISO 9001** is a worldwide quality management system (QMS) standard.
- An **Internal Audit** is a documented review of a group or particular organization to verify if the activities that affect quality and/or environmental management are documented and that the system is effective.
- Auditors are designated personnel who have satisfied the minimum requirements of formal training to the ISO 9001 and/or ISO 14001 Management Standards. In order to be a lead auditor, personnel must have completed two (2) previous audits.

Referenced documents

- ISO 14001, Element 4.5.4, EMS Audit
- ISO 9001, Element 4.17, Internal Quality Audits
- QAP #12345, Quality Manual
- QAP #56789, Environmental Manual
- SOP #24680, Corrective and Prevention Action

EH&S precautions   During all internal audits, all safety precautions (i.e., PPE) shall be adhered to.

Precedence   If there is a conflict between this procedure and the International Standards, the latter shall take precedence.

Responsibilities

- The Quality Assurance Manager and the Environmental Manager shall jointly be responsible for organizing and planning internal audits, reviewing audit findings, participating in the classification of nonconformances, as well as tracking actions identified in the audit until such time as they are corrected.

- Audits may be comprised of trained members from other functional areas as long as they are not directly responsible for the area being audited.
- The Quality Assurance Department shall be responsible for filing audit reports and any response reports.
- The organization or area being audited shall be responsible for responding to the audit findings with a corrective action plan and within the established timeframe.

Procedure

Audit Plan

- Quality Assurance and/or Environmental Management prepares an Audit Plan

Schedule

- The approximate months in which audits will occur shall be established, which includes the areas to be audited.

Auditing Criteria

- Quality audits shall take into consideration the results of previous audits.
- Environmental audits shall consider environmental importance a criteria for frequency, as well as the results of previous audits.

Frequency of Audits

- The standard frequency for each area shall be twice a year.
- Exceptions:
  — Areas with ≤ 1 Class 2 Corrective Actions will be skipped the next time they would normally be audited.
  — Areas with ≥ 5 Class 2 Corrective Actions will be audited more frequently (no more than three times per year).
  — Areas with two to four Corrective Actions may be skipped the next time they would normally be audited depending on the severity of the corrective actions.
- Other internal audits conducted by Corporate will be conducted on an annual basis.

Planning

- The area to be audited shall be notified that an audit is scheduled and to arrange for escort personnel to be available.
- A preaudit meeting is held with the area manager(s).
- Auditors should request copies of any procedures, copies of previous audits, any outstanding corrective actions or any other supporting documentation.
- Auditors shall obtain copies of the internal audit checklist.

Conducting the Audit

- Obtain a verbal explanation of the process or area being audited.
- Verify which aspects of the process affect quality and the environment are written into the procedure(s) and ensure they agree with the verbal explanation.
- Verify what records exist to determine if the process was performed according to the procedures.

Post Audit

- A post-audit meeting is held to confirm that the findings are accurate and to review documents and records which may have been unavailable at the time of the audit. Nonconformances are discussed at this point.

Audit Report

- The audit team issues a Quality Assurance/Environmental Audit Report within 7 calendar days of the audit.
- The audit report is distributed to the area manager and to any other appropriate personnel. An additional copy is filed.

Corrective Actions

- Corrective actions are classified as to risk by the Quality and/or Environmental Manager(s).
- A corrective action form is issued.

Response by Auditee

- The audited area prepares a memo within 14 calendar days which addresses the nonconformances and presents any corrective actions planned. The responsible person(s) and the estimated completion timeframe are also included.
- Corrective actions are to be completed per the timeframes.

19.5.1 SCHEDULE

When establishing an auditing schedule, you must take into consideration not only what areas are to be audited, but also what elements of the ISO Standards are to be included. It is highly recommended early on in the process that each area be audited much more frequently and then gradually scheduled on a less frequent basis after two or three years.

Table 19.2 shows a schedule for auditing the ISO 14001 Elements. A similar schedule can also be assembled for a combined ISO 9001 and ISO 14001 audit. The schedule shows that some of the elements are audited on a much more frequent basis, while a few are audited on an annual basis. In contrast, Table 19.3 shows which specific areas or departments will be audited and during which month of the year. The last line of the schedule is marked "Systems" and refers to a system-wide audit to be conducted during the months of November and December (Q2 on the fiscal year).

TABLE 19.2
Internal Audit Schedule

Element	Description	FY98 Q1	FY98 Q2	FY98 Q3	FY98 Q4	FY99 Q1	FY99 Q2	FY99 Q3	FY99 Q4	FY00 Q1	FY00 Q2	FY00 Q3	FY00 Q4
4.2	Environmental Policy	•	•	•	•	•	•	•	•	•	•	•	•
4.3.1	Environmental Aspects	•	•	•	•	•	•	•	•	•	•	•	•
4.3.2	Legal and Other Requirements		X				X				•		
4.3.3	Objectives and Targets	•	•	•	•	•	•	•	•	•	•	•	•
4.3.4	Environmental Management Programs	•	•	•	•	•	•	•	•	•	•	•	•
4.4.1	Structure and Responsibility		X				X				X		
4.4.2	Training, Awareness, & Competence	•	•	•	•	•	•	•	•	•	•	•	•
4.4.3	Communications	•	•	•	•	•	•	•	•	•	•	•	•
4.4.4	EMS Documentation		X				X				•		
4.4.5	Document Control		X				X				•		
4.4.6	Operational Control	•	•	•	•	•	•	•	•	•	•	•	•
4.4.7	Emergency Preparedness & Response	•	•	•	•	•	•	•	•	•	•	•	•
4.5.1	Monitoring & Measurement		X				X				X		•
4.5.2	Nonconformance and Corrective and Preventive Action	•	•	•	•	•	•	•	•	•	•	•	•
4.5.3	Records		X				X				X		
4.5.4	EMS Audit		X				X				X		
4.6	Management Review		X				X				X		

• *Specific* elements covered each month during department internal audits/EHS Inspections
X *Systemic* elements covered during annual Raychem audits (in addition to other elements)

TABLE 19.3
Department ISO Audit Schedule

Dept #	FY98												FY99					
Area	Jul	Aug	Sep	Oct	Nov	Dec	Jan	Feb	Mar	Apr	May	Jun	Jul	Aug	Sep	Oct	Nov	Dec
Assy 1			•												•			
Assy 2							•										•	
Tech.										•								
Eng.											•							
MC				•												•		
QC								•										
CSC	•												•					
EHS									•									
Lab 1		•												•				
Lab 2												•						
Systems					•	•												•

Besides scheduling areas and elements, it is also important to have identified auditing teams and an auditing schedule for them, as well. Table 19.3 can be modified to include the audit teams by including some team "code" in the respective boxes.

19.6   WHAT AUDITORS WILL LOOK FOR

In this Section I am referring to a third party auditor of an ISO certifying organization. The auditor will be concerned about the basic methodology of your internal auditing system and the scheduling process. In conjunction with this element, the auditor will want to evaluate your corrective action process. The last concern will be in relation to the training of the audit members and the assurance that they are qualified to understand and address the specific ISO requirements.

Part VI

Management Review

20 Management Review

20.1 INTRODUCTION

We now come to the penultimate chapter which deals with the subject of management's responsibility to review the information, records, data, and documentation against the ISO 9001 and ISO 14001 requirements. A great deal of effort will be spent establishing and implementing your integrated program, but unless appropriate senior management is involved in the process, the system is doomed to failure.

It is senior management which will need to appraise the progress of the system and assess *how* well it is working, *where* it is complying or not complying with the standards, *what* needs to be done to continually improve the process, *who* will need to be involved (i.e., resources), and *when* (the time table) corrective actions will need to be completed. This chapter is intended to provide you with an example of a framework to assist management in understanding its responsibilities in ensuring your company's compliance to ISO 9001 and ISO 14001.

20.2 COMPARISON OF THE STANDARDS

Table 20.1 shows the comparison of the standards and will be used to structure the rest of the chapter. As you evaluate this table, you will notice that ISO 14001 has much more specific requirements and states the potential need to address changes — all with the goal of ensuring compliance with the policy commitment to continual improvement. Although I did not include information from ISO 9004, I think it is important to briefly look at what it states in order that we have more information to use when integrating the two systems. In Element 5.5 of ISO 9004, Review and Evaluation of the Quality System, it states:

> The organization's management should provide for independent review and evaluation of the quality system at defined intervals. The reviews of the quality policy and objectives should be carried out by top management, and the review of supporting activities should be carried out by management with executive responsibilities for quality and other appropriate members of management, utilizing competent independent personnel as decided on by the management.
>
> Reviews should consist of well-structured and comprehensive evaluations which include:
>
> (a) results from internal audits centered on various elements of the quality system;
> (b) the overall effectiveness in satisfying the guidance of this part of ISO 9004 and the organization's stated quality policy and objectives;

TABLE 20.1

Correlation of "Management Review" Requirements

ISO 9001		ISO 14001	
Section	**Requirements**	**Section**	**Requirements**
The supplier's management with executive responsibility shall ...		The organization's top management shall, at intervals it determines,	
4.1.3	review the quality system at defined intervals	4.6 (i)	review the environmental management system,
sufficient to ensure its ...		to ensure its ...	
4.1.3	continuing suitability and	4.6 (i)	continuing suitability,
		4.6 (i)	adequacy and
4.1.3	effectiveness	4.6 (i)	effectiveness.
		The management review process shall ensure that ...	
		4.6 (i)	the necessary information is collected to allow management to carry out this evaluation.
Records of such reviews shall be ...		This review shall be ...	
4.1.3	maintained.	4.6 (i)	documented.
in satisfying the requirements of		The management review shall address the possible need for changes to ...	
4.1.3	stated quality policy and	4.6 (ii)	the policy,
4.1.3	objectives	4.6 (ii)	objectives and
4.1.3	International Standards	4.6 (ii)	other elements of the environmental management system,
		in light of ...	
		4.6 (ii)	environmental management system audit results,
		4.6 (ii)	changing circumstances and
		4.6 (ii)	the commitment to continual improvement

(c) considerations for updating the quality system in relation to changes brought about by new technologies, quality concepts, market strategies, and social or environmental conditions.

Observations, conclusions and recommendations reached as a result of review and evaluation should be documented for necessary action.

As you can see here, the ISO 9004 document is much more in line with the ISO 14001 requirements.

TABLE 20.2
List of Review Requirements

ISO 9004		ISO 14001	
Section	**Document**	**Section**	**Document**
4.2 and 5.5	Quality Policy	4.2	Environmental Policy
5.5	Supporting Activities	4.3.1	Environmental Aspects
4.3.1 and 5.5	Objectives	4.3.2	Legal Requirements
5.2.4	Resources	4.3.3	Objectives and Targets
5.4.5	Follow-up Action	4.3.4	EM Programs
5.4.4 and 5.5 (a)	Auditing	4.4.1	Resources
5.3.3	Quality Plan	4.4.3	External Communication
6.1	Financial considerations	4.4.7	Emergency Preparedness and Response Plan
7.1	Marketing requirement	4.5.1	Monitoring and Measurement Data
8.4	Design Reviews	4.5.2	Nonconformances and Corrective Actions
9.1	Purchasing contracts	4.5.2	Preventive Activities
15.1	Corrective Actions	4.5.4	Audits

20.3   WHAT SHOULD BE REVIEWED?

Before we begin to construct a procedure and a framework for management to use, it is first important to understand exactly what must be reviewed. This will require a detailed evaluation of the two ISO Standards and the compilation of a list of information that can be used to construct a procedure. Since ISO 9004 is much more detailed, it will be used to list the review requirements in Table 20.2.

It is not necessary for every single aspect of your operation to be reviewed by senior management. Unless there is the potential for a major nonconformance, the day-to-day issues of your process need not be brought to their attention. If your system is functioning properly, management reviews need only take place at a minimum of once a month. The next two sections will detail procedures which can be adopted to assist management in the review process.

In Document Control systems, documents are typically classified according to four (4) levels: standards and manuals are classified *Level 1* documents; documents which describe overall QA and EM program requirements are listed as *Level 2*; area, process, or work instruction documents are classified as *Level 3*: and data and other forms of metrics are classified as *Level 4*. The first procedure to be presented is classified as a Level 2 Document and defines the overall management review process and how the quality and environmental policies are to be implemented. The second procedure is classified as a Level 3 Document and is specific to the Environmental (Health and Safety) Management Committee. The Procedural Document contains detailed requirements for ISO 14001. For both procedures I will continue to use the standard procedural format employed throughout this book.

20.4 MANAGEMENT REVIEW PROCESS PROCEDURE

This Level 2 procedure is a document that describes the processes by which senior management will review the organization's quality and environmental management systems and ensure that appropriate changes will be made to the system to maintain compliance with the standards.

QUALITY ASSURANCE PROCEDURE FOR MANAGEMENT REVIEW

Purpose   This procedure defines the processes by which the company senior management will review the Quality and Environmental Management Systems in order to ensure and verify their continuing suitability and effectiveness. It will also describe the process for internally and externally deploying the Quality and Environmental Policies.

Scope   This procedure applies to the company's management review of the Quality and Environmental Management Systems as they are described in the Operating Manual.

Definition of terms

- **ISO 14001** is the worldwide environmental management system (EMS) standard.
- **ISO 9001** is a worldwide quality management system (QMS) standard.
- An **Internal Audit** is a documented review of a group or particular organization to verify if the activities that affect quality and/or environmental management are documented and that the system is effective.
- **Management Review** is a periodic, defined and systematic method of evaluating the continuing suitability and effectiveness of the Quality Management System by Senior Management and of the Environmental Management System by the Environmental Management Committee.
- The **Quality Improvement Committee** is the company's organization for reviewing all aspects of the quality management system.
- The **Environmental Management Committee** is the company's organization for reviewing all aspects of the environmental management system.
- The **Environmental Management System** is a formal process based on written procedures, federal, state and local regulations, and company standards that are used as a means to ensure the protection of the environment, people, property and the community from any adverse effects.
- The **Quality Management System** is a formal process based on written procedures used to ensure product is consistently conforming to specified requirements.

Referenced documents

- ISO 14001, Element 4.6, Management Review
- ISO 9001, Element 4.1.3, Management Review
- QAP #12345, Quality Manual

- QAP #56789, Environmental Manual
- SOP #24680, Corrective and Prevention Action
- QAP #13579, Customer Complaints
- SOP #34589, Internal Audits
- SOP #23759, Environmental Management Committee

EH&S precautions   NA

Precedence   If there is a conflict between this procedure and other management review processes, this procedure shall take precedence.

Responsibilities

Quality Improvement Team

- Reviewing the Quality Management System at defined intervals in order to ensure its continuing suitability and effectiveness to the ISO 9001 Quality Management Standards and against customer-specific quality requirements.

Quality Assurance Manager

- Ensuring that Management Reviews of the Quality Management System are scheduled and conducted.
- Reporting on the results of the Quality Improvement Team.

Environmental Management Committee

- Reviewing the Environmental Management System at defined intervals in order to ensure its continuing suitability and effectiveness to the ISO 14001 Environmental Management Standards and against regulatory requirements.

Environmental Manager

- Ensuring that the Environmental Management Committee meetings are scheduled and conducted.
- Reporting on the results of the Environmental Management Committee.

Procedure

A. On a monthly basis, the Quality Improvement Team meets to review the performance of quality indicators which includes the following:
   - The results of internal audits
   - Customer complaints (i.e., product returns, etc.)
   - Open corrective actions
   - The results of external and customer audits
   - Delivery performance
   - Training
   - Product Development
   - Discrepant material/product
   - Supplier performance

B. On a monthly basis, the Environmental Management Committee meets to review the performance of environmental indicators as defined in SOP #23759.

C. On a quarterly basis, the Management Review evaluates progress against major company objectives that are associated with Key Indicators. The result of this review is to consider, if required, the establishment of new or modified objectives.

D. The company's Quality Policy is made available within the organization through the area and department-specific Continuous Improvement Teams. Each department manager is responsible for ensuring that the policy is understood, implemented and maintained at all levels of their respective departments. Training records are maintained.

20.5   ENVIRONMENTAL MANAGEMENT COMMITTEE

As mentioned previously, companyies (primarily larger organizations) may have a separate committee to deal with environmental (health and safety) issues and be responsible for the management review requirements of ISO 14001. The procedure presented in this section is intended to focus strictly on such a committee and is written in such a manner as to address specific ISO 14001 requirements. This document is a Level 3 procedure and, you will note, was referenced in the Management Review Process Procedure in the previous section.

STANDARD OPERATING PROCEDURE #23759
ENVIRONMENTAL MANAGEMENT COMMITTEE

Purpose   The purpose of this procedure is to define the function and responsibilities of the Environmental Management Committee (EMC).

Scope   This procedure covers all of the requirements governing the function of the Environmental Management Committee as they relate to ISO 14001 Environmental Management Standards and federal, state and local environmental, health and safety regulations.

Definition of terms

- **ISO 14001** is the worldwide environmental management system (EMS) standard.
- **Environmental Aspects** are those particular functions or processes which may have a potential impact on the environment, people, and/or property.
- **Environmental Impacts** are the effects which a particular aspect may have on the environment, people, and/or property.
- **Significant Aspects** are those particular Environmental Aspects defined above which may have a significant impact as defined by the Risk Assessment process. The Risk Assessment formula (below) defines "significant" as a level of risk ≥ 50 on the numerical scale.

- **Risk Assessment** is a formal process where a formula (severity x probability) is utilized to assess and define the level of risk or impact associated with a particular environmental aspect.

Referenced documents

- ISO 14001, Element 4.6, Management Review
- QAP #56789, Environmental Manual
- SOP #24680, Corrective and Prevention Action
- QAP #13579, Customer Complaints
- SOP #34589, Internal Audits
- SOP #23759, Environmental Management Committee
- SOP #87363, Objectives, Targets and Programs
- SOP #59274, Ensuring Legal and Standards Compliance

EH&S precautions   NA

Precedence   The functions of the Environmental Management Committee described in this procedure shall only be amended in the event of changes in regulatory legislation, the ISO 14001 Environmental Management Standards, and/or the internal environmental policy and its accompanying standards.

Responsibilities

Senior Manager

- Ensure that an Environmental Manager is appointed to facilitate the monthly meetings;
- Attend each monthly as his/her schedule will allow;
- Communicate appropriate information from the meeting to his/her staff; and
- Review all information reported by the Environmental Manager in order to assist in rendering a decision on issues or matters being presented.

Operations Managers

- Attend each monthly as his/her schedule will allow;
- Communicate appropriate information from the meeting to his/her staff;
- Review all information reported by the Environmental Manager in order to assist in rendering a decision on issues or matters being presented; and
- Designate a substitute to attend the meeting in the event he/she is unable to attend.

Production Managers

- Attend each monthly as his/her schedule will allow;
- Communicate appropriate information from the meeting to his/her staff;
- Review all information reported by the Environmental Manager in order to assist in rendering a decision on issues or matters being presented; and
- Designate a substitute to attend the meeting in the event he/she is unable to attend.

Supervisors

- Attend each monthly as his/her schedule will allow;
- Communicate appropriate information from the meeting to his/her staff;
- Review all information reported by the Environmental Manager in order to assist in rendering a decision on issues or matters being presented; and
- Designate a substitute to attend the meeting in the event he/she is unable to attend.

Sales/Marketing/Customer Service Managers

- Attend each monthly as his/her schedule will allow;
- Communicate appropriate information from the meeting to his/her staff;
- Review all information reported by the Environmental Manager in order to assist in rendering a decision on issues or matters being presented; and
- Designate a substitute to attend the meeting in the event he/she is unable to attend.

Develop/Technical Managers

- Attend each monthly as his/her schedule will allow;
- Communicate appropriate information from the meeting to his/her staff;
- Review all information reported by the Environmental Manager in order to assist in rendering a decision on issues or matters being presented; and
- Designate a substitute to attend the meeting in the event he/she is unable to attend.

Environmental Manager

- Schedule all monthly meetings according to the agreed upon time and dates established by the members of the Environmental Management Committee;
- Assemble information, data, etc. required for the meeting and prepare the agenda for distribution before the meeting;
- Review minutes of the meeting which includes the listed agenda and all comments/considerations, assignment of corrective actions, and final decisions;
- Maintain records of all documentation related to the function of the Environmental Management Committee;
- Ensuring that the Environmental Management Committee meetings are scheduled and conducted.
- Reporting on the results of the Environmental Management Committee.

Procedure

A. Identify members of the Environmental Management Committee which shall consist of the following representation:
   - Senior Operating Manager
   - Operations Manager
   - Production Manager

- Supervisor
- Develop/Technical Manager
- Marketing/Sales/Customer Service Manager
- Environmental Manager
- Quality Assurance Manager (optional)

B. Establish regular meeting schedule.

C. In accordance with the requirements of the company and other potentially-related codes, standards and regulations, establish the organization's Environmental Policy.
- The Environmental Policy shall be evaluated and amended as appropriate at the beginning of each calendar year.
- The Environmental Policy shall be publicly available in the following ways:
  — posted in the main lobby of all company buildings
  — posted throughout each building and on various bulletin boards
  — available to external interested parties (agencies, customers, stockholders, neighbors, etc.) upon request and through the issuance of the annual environmental report
  — included in new hire orientation packages
  — available to all employees in either hard copy or electronically

D. Identify all regulatory, codes, standards and other requirements which the members must evaluate or establish during the course of each year as they may apply to or may impact environmental objectives. The requirements may include, but are not limited to the following:
- Corporate Environmental Policy and accompanying Standards;
- ISO 14001;
- U.S. Federal and State Environmental Regulations;
- Local Environmental Codes; and
- Uniform Fire Code

E. Objectives and Targets
- At the beginning of each year the members of the Environmental Management Committee shall assist their respective departments for the purpose of evaluating and establishing objectives and targets that are based primarily on the significant environmental aspects identified by the Environmental Department. Departments are not restricted to those that are "significant." The review shall consider the suitability, adequacy and effectiveness of the targets and objectives;
- Establish the overall objectives and targets and evaluate whether they are in compliance with the Environmental Policy;
- Review the objectives and targets on a monthly basis in order to evaluate their progress and amend them if the progress is determined to be impacting the company's compliance to those listed in Section D above;

- In the event of a new product or process change, evaluate objectives and targets to determine if any impact will result. Amend the objectives and targets as appropriate.

F. Environmental Aspects
  - The Environmental Management Committee shall review the organization's environmental aspects on a regular basis; and
  - Shall take into consideration any processes for externally communicating any significant aspects.

G. Management Review
  - One of the primary functions of the Environmental Management Committee is to review at a minimum of once per year the Environmental Management System which will include all of the following:
    — Corporate and Company Environmental Policies;
    — Results of any internal audits which focus on compliance with ISO 14001 and/or Corporate audits and evaluations; and
    — Results of any external audits which arise through ISO 14001 certification, regulatory inspections and customers.
  - As a result of the audits, identify any corrective actions and monitor progress through the corrective action program.

20.6   WHAT AUDITORS WILL LOOK FOR

During an ISO audit, the auditor will look for documented evidence of the following: (a) the policy been reviewed and/or modified to reflect any changes in the operation internally or have been influenced by external concerns; (b) a consideration of methods to externally communicate its significant environmental aspects; (c) a review of the organization's objectives and targets and has this review taken into consideration other requirements and its significant aspects in establishing them; (d) a consideration of methods to make the environmental policy available externally; (e) reviews of internal and external audits; (f) the reviews took place (meeting minutes); and (g) the reviews address the company's commitment to continual improvement and the prevention of pollution.

Part VII

Conclusion

21 The Future

21.1 ISO HEALTH AND SAFETY STANDARDS

In early 1995, even before the final draft of ISO 14001 had been printed and issued, inquiries began to circulate concerning the possible need for a corresponding international health and safety standard. At first glance this seems to make a lot of sense and the implementation of such a standard could be done very easily with an environmental management system already in place. Many companyies realize that a single management standard to deal with quality, environmental and health and safety would be both cost effective and much more efficient.

In 1994, the British Standards Institute (BSI) developed BS 8750, *Guide to Occupational Health and Safety Management*, which is expected to be a potential precursor to any ISO Health and Safety Standards. One of the unique features of the standard is that it also provides guidance on integrating occupational health and safety management systems into a company's overall management system.

Surveys, however, have indicated that there is little support for the development of health and safety standards. Many of the industrialized nations feel that they have substantial regulations in place and do not want to be burdened with yet another international standard.

21.2 RIO +5

In June 1997, five years after the 1992 Earth Summit in Rio de Janeiro, a follow-up summit was held to evaluate the progress of the Agenda 21 initiatives. The Rio +5 summit brought together participants form several sectors, agencies and societies with the intent to answer some of the following questions:

- Assess the progress of Sustainable Development in terms of financial requirements and other economic systems.
- Share the best Sustainable Development practices.
- Build a better Sustainable Development infrastructure.
- Evaluate the gaps between Agenda 21 and reality.

The results of the summit have indicated that there has been very little progress on reducing environmental impacts and implementing Sustainable Development. Several key issues identified at the summit were: (a) the growing presence of toxic chemicals in the environment; (b) a growing scarcity of fresh water and the accompanying loss of productive farmland; (c) the continued destruction of forest; (d) growing marine pollution and its accompanying reduction in viable fisheries; and (e) the increase in global warming from heat trapping gases such as carbon dioxide.

Although it has been generally agreed that the summit made virtually no progress, participants did agree to a formal document regarding the preserving of foresty and cutting back on "greenhouse" gases. The result of this was a convention on global warming in December 1997, held in Kyoto, Japan. At this conference many nations signed an agreement to mandatory cutbacks in the emissions of "greenhouse gases" such as carbon dioxide and reached a consensus on specific timetables. Many less industrialized nations and the EU would like to see a 15% reduction by 2010, but the U.S. automakers have indicated that such a timeframe will throw the world into a recession. Additionally, leading up to the conference, the United States made strong statements concerning the future role of less-industrialized nations on global warming — in a few decades, the less industrialized nations will be impacting global warming at a much greater rate and they must take responsibility now, not later.

It will be interesting to see the results at "Rio +10."

21.3   FINAL COMMENTS

Although the ISO 14001 Environmental Management Standards have not really taken hold in the United States, there are a few "signs" which may lead one to speculate on its progress. In late 1995, for instance, the Big Three U.S. Automakers speculated that they might go beyond the ISO 14001 requirements by developing a QS-14000 program and requiring their suppliers to have a joint QS-9000 and ISO 14001 third party certification. Representatives from the automakers have cited increasing environmental regulations as the reason for this speculation.

On the government level, in 1995 President Bill Clinton signed the National Technology Transfer and Advancement Act into law which requires federal agencies to use existing technical standards (i.e., ANSI, ASTM, ASQC, and ASME for instance). For environmental compliance it is expected that federal agencies and facilities may look at the ISO 14000 series of standards. Companies supplying any materials and services to the federal government may ultimately also be required to implement the standards, as well.

Near the end of 1997, there were less than 50 companies in the United States certified to ISO 14001. In Europe and Asia, the list is in the hundreds. Even though several large U.S. firms were actively involved in the development of the standard, there has been relatively little progress in its acceptance in the United States. Much like the ISO 9001 Standards, many U.S. companies will feel the pressure to become certified to ISO 14001 from their customers, especially those in Europe and Asia.

If you are contemplating ISO 14001, it is highly recommended that your company seriously evaluate its implementation by integrating it into your existing quality management system. If for some reason you have not adopted a quality standard and you are strongly considering ISO 14001, this book may assist you to some degree in improving your quality management system as you implement the environmental management standard. The integration process has proven to be a very effective and cost efficient way to develop one management standard.

References

1. Abhay K. Bhusan and James C. MacKenzie, "Environmental Leadership Plus Total Quality Management Equals Continuous Improvement," in *Understanding Total Quality Environmental Management*, Executive Enterprises Publications Co., 1992
2. CEEM Information Services, *Int. Env. Syst. Update*, July 1995, p. 12
3. CEEM Information Services, *Int. Env. Syst. Update*, April 1995, p. 11
4. CEEM Information Services, *Int. Env. Syst. Update*, December 1995, p. 22
5. CEEM Information Services, *Int. Env. Syst. Update*, January 1996, p. 1
6. CEEM Information Services, *Int. Env. Syst. Update*, September 1996, p. 1

Appendix A
List of Abbreviations
and Acronyms

ANSI	American National Standards Institute
API	American Petroleum Institute
ASQC	American Society of Quality Control
BSI	British Standards Institute
CAA	Clean Air Act (U.S.)
CEN	Comitee Europeen de Normalization or Committee for European Standardization
CENELEC	Comitee Europeen de Normalization Electrotechnique or European Committee for Electrotechnical Standardization
CERCLA	Comprehensive Environmental Response Compensation and Liability Act (Superfund)
CMA	Chemical Manufacturers Association
CSA	Canadian Standards Association
CWA	Clean Water Act (U.S.)
DFE	Design for Environment
DFM	Design for Manufacturability
EC	European Commission
EH&S	Environmental, Health, and Safety
EIMS	Environmental Information Management System
EMAS	Eco-Management Audit Scheme
EMS	Environmental Management System
EPA	Environmental Protection Agency (U.S.)
EU	European Union
FIFRA	Federal Insecticide, Fungicide, and Rodenticide Act (U.S.)
GATT	General Agreement on Trades and Tariffs
GEMI	Global Environmental Management Initiative
HZPO	Hazard Potential
ICC	International Chamber of Commerce
IEC	International Electrotechnical Commission
IISD	International Institute for Sustainable Development
INE	National Institute of Ecology (Mexico)
ISO	International Standardization Organization
JIS	Japanese Industrial Standards
JUSE	Japan Union of Scientists and Engineers

MITI	Ministry of International Trade and Industry (Japan)
NAFTA	North American Free Trade Agreement
NEPA	National Environmental Policy Act (U.S.)
OSHA	Occupational Safety and Health Administration (U.S.)
PROFEPA	Procuraduria Federal de Proteccion al Ambiente (Mexico)
QA	Quality Assurance
QMI	Quality Management Institute of Canada
QMS	Quality Management System
QS-9000	Quality Systems Requirements (U.S. Automotive)
QSR	Quality Systems Review (Motorola Corporation)
RCRA	Resource Conservation Recovery Act (U.S.)
SAGE	Strategic Advisory Group for the Environment
SEMARNAP	Ministry of Environment, Natural Resources, and Fisheries (Mexico)
SME	Small to Medium-sized Enterprise
SMOP	Synthetic Minor Operating Permit
SQC	Statistical Quality Control
STEP	Strategies for Today's Environmental Partnership
TQC	Total Quality Control
TQEM	Total Quality Environmental Management
TQM	Total Quality Management
TSCA	Toxic Substances Control Act (U.S.)
UNCED	United Nations Conference on Environment and Development

Appendix B
Comparison of ISO 9001 and ISO 14001

ISO 9001 — Quality Management System (QMS)		ISO 14001 — Environmental Management System (EMS)	
Element	**Description**	**Element**	**Description**
1	Scope	1	Scope
2	Normative References	2	Normative References
3	Definitions	3	Definitions
4	Quality System Requirements	4	Environmental Management System Requirements
4.1	Management Responsibility	4.3.4(a) / 4.4.1	Environmental Management Programs / Structure and Responsibility
4.1.1	Quality Policy	4.2 / 4.3.3	Environmental Policy / Objectives and Targets
4.1.2	Organization	4.4.1	Structure and Responsibility
4.1.2.1	Responsibility and Authority	4.4.1 / 4.5.2	Structure and Responsibility / Nonconformance and Corrective/Preventive Action
4.1.2.2	Resources	4.4.1(ii)	Structure and Responsibility
4.1.2.3	Management Representative	4.4.1(iii)	Structure and Responsibility
4.1.3	Management Review	4.6	Management Review
4.2	Quality System		
4.2.1	General	4.4.4	EMS Documentation
4.2.2	Quality System Procedures	4.4.6 (b)	Operational Control
4.2.3	Quality Planning	4.3	Planning
4.3	Contract Review		
4.3.1	General	4.3.1 / 4.4.6(c)	Environmental Aspects / Operational Control
4.3.2	Review		

ISO 9001 — Quality Management System (QMS)			ISO 14001 — Environmental Management System (EMS)	
Element		Description	Element	Description
	4.3.3	Amendment to a Contract		
	4.3.4	Records	4.5.3	Records
4.4		Design Control		
	4.4.1	General	4.3	Planning
	4.4.2	Design and Development Planning		
	4.4.3	Organizational and Technical Interfaces	4.4.1(i) 4.4.3(a)	Structure and Responsibility Communication
	4.4.4	Design Input	4.3.1 4.3.2	Environmental Aspects Legal and Other Requirements
	4.4.5	Design Output	4.3.1 4.3.2	Environmental Aspects Legal and Other Requirements
	4.4.6	Design Review	4.5.3 4.6	Records Management Review
	4.4.7	Design Verification		
	4.4.8	Design Validation		
	4.4.9	Design Changes	4.3.1 4.3.2	Environmental Aspects Legal and Other Requirements
4.5		Document and Data Control		
	4.5.1	General	4.4.5 4.5.3	Document Control Records
	4.5.2	Document and Data Approval and Issue	4.4.5	Document Control
	4.5.3	Document and Data Changes	4.4.5	Document Control

ISO 9001 — Quality Management System (QMS)		ISO 14001 — Environmental Management System (EMS)	
Element	**Description**	**Element**	**Description**
4.6	Purchasing		
4.6.1	General	4.3.1 4.4.6	Environmental Aspects Operational Control
4.6.2	Evaluation of Subcontractors	4.3.1 4.4.6(c)	Environmental Aspects Operational Control
4.6.3	Purchasing Data	4.5.3(ii)	Records
4.6.4	Verification of Purchased Product		
4.6.4.1	Supplier Verification at Subcontractor Premises		
4.6.4.2	Customer Verification of Subcontractor Product	4.3.1 4.4.6(c)	Environmental Aspects Operational Control
4.7	Control of Customer-Supplied Product	4.3.1 4.4.6(c)	Environmental Aspects Operational Control
4.8	Product Verification and Traceability	4.3.1 4.4.6(b)	Environmental Aspects Operational Control
4.9	Process Control	4.4.6	Operational Control
4.10	Inspection and Testing		
4.10.1	General		
4.10.2	Receiving Inspection and Testing		
4.10.3	In-Process Inspection and Testing		
4.10.4	Final Inspection and Testing		
4.10.5	Inspection and Test Records		

ISO 9001 — Quality Management System (QMS)			ISO 14001 — Environmental Management System (EMS)	
Element		Description	Element	Description
4.11		Control of Inspection, Measuring, and Test Equip		
	4.11.1	General	4.3.6 4.5.1(ii)	Operational Control Monitoring and Measurement
	4.11.2	Control Procedure	4.5.1(i)	Monitoring and Measurement
4.12		Inspection and Test Status		
4.13		Control of Nonconforming Product		
	4.13.1	General	4.3.1 4.4.6(a)	Environmental Aspects Operational Control
	4.13.2	Review/Disposition of Nonconforming Product		
4.14		Corrective and Preventive Action	4.5	Checking and Corrective Action
	4.14.1	General	4.5.2	Nonconformance and Corrective/Preventive Action
	4.14.2	Corrective Action	4.4.3(b) 4.5.2	Communication Nonconformance and Corrective/Preventive Action
	4.14.3	Preventive Action	4.5.2 4.6	Nonconformance and Corrective/Preventive Action Management Review
4.15		Handling, Storage, Packaging, Preservation, and Delivery		
	4.15.1	General	4.3.1 4.4.6	Environmental Aspects Operational Control
	4.15.2	Handling	4.4.2	Training, Awareness, and Competence
	4.15.3	Storage	4.4.6	Operational Control

ISO 9001 — Quality Management System (QMS)		ISO 14001 — Environmental Management System (EMS)	
Element	**Description**	**Element**	**Description**
4.15.4	Packaging	4.3.1 4.3.2	Environmental Aspects Legal and Other Requirements
4.15.5	Preservation		
4.15.6	Delivery		
4.16	Control of Quality Records	4.4.5 4.5.3	Document Control Records
4.17	Internal Quality Audits	4.5.4	EMS Audit
4.18	Training	4.4.2 4.4.7	Training, Awareness, and Competence Emergency Preparedness and Response
4.19	Servicing		
4.20	Statistical Techniques		
4.20.1	Identification of Need	4.5.1	Monitoring and Measurement
4.20.2	Procedures		

(i, ii, iii, ...) denotes specific paragraph under subelement

(a, b, c, .) denotes specific subsection under subelement

Appendix C
Related Sections of ISO 14001, ISO 9001, BS 7750, and EMAS

ISO 14001	ISO 9001	BS 7750	EMAS
SCOPE: Specifies EMS requirements without specific performance criteria. Company formulates policy and objectives and considers legal and other requirements. Must identify significant impacts.	SCOPE: Specifies quality system requirements for suppliers who have the capability to design and provide specific customer products.	SCOPE: Specifies development, implementation and maintenance of EMS, without specific performance criteria. Company formulates policy and objectives and must identify significant environmental effects.	EMAR Article 1: EMAS intended to evaluate and improve environmental performance of a company performing industrial activities; Participation is voluntary
4.1 Requirements: An organization must implement and maintain an EMS to be certified.	4.0 Requirements: (Contained in all of Section 4)	4.1 Requirements: Organization must implement and maintain an EMS to be certified. Requires documented system and implemented procedures.	EMAR Article 3: Registration requires a verified policy program, an EMS review or audit procedure, and environmental statement.
4.2 Environmental Policy: Management shall define and ensure policy is appropriate to business, commit to continual improvement and prevention of pollution; public availability and communication; frame-work for policy review, maintenance, documentation, and implementation	4.1.1 Quality Policy: Supplier's management shall define, document, implement, and maintain its policy and objectives.	4.2 Environmental Policy: Must define policy and ensure relevancy; ensure it is communicated, implemented, main-tained, publicly available, and continually improved; must publish objectives, define environmental activities, and show how objectives will be made public	Annex I.A.1, 2, 3, 4: Company policy shall be in writing, reviewed on a periodic basis, revised, and communicated to employees and publicly available; must specify objectives.
4.3 Planning	4.2.3 Quality Planning	No Similar Title	No Similar Title
4.3.1 Environmental Aspects must be considered when setting objectives; shall establish procedures to identify environmental impacts.	Supplier shall define and document how the requirements for quality shall be met.	4.4 Environmental Effects: 4.4.1 Effects must be communicated; 4.4.2 Compile evaluation register: effects must be identified and register maintained	Annex I.B.3: Environmental effects: Company activities must be evaluated and significant effects listed in a register.
4.3.2 Legal and Other Requirements shall be identified and considered as an environmental aspect.	4.4.4 Design Input: Requirements shall include applicable statutory and regulatory requirements	4.4.3 Legislative, regulatory, and other policy requirements shall be established and maintained in a register.	Annex I.B.3: Legislative and other policy requirements shall be est. and maintained in a register.

ISO 14001	ISO 9001	BS 7750	EMAS
4.3.3 Objectives and targets shall be consistent with environmental policy, including commitment to prevention of pollution	4.1.1 Quality Policy shall include quality objectives and commitments	4.5 Environmental objectives and targets shall be consistent with environmental policy and include commitment to continual improvement.	Annex I.A.4: Environmental objectives shall be consistent with environmental policy and include commitment to continual improvement.
4.3.4 Environmental management program(s) shall include designation of responsibility and timeframe for achieving objectives and targets	4.2.2 Quality system procedures shall prepare and effectively implement quality system and documented procedures	4.6 Environmental management program shall include designation of responsibility and the means for achieving targets.	Annex I.A.5: Environmental program shall include designation of responsibility and the means by which objectives will be achieved.
4.4 Implementation and Operation	4.1.2 Organization	4.3 Organization and Personnel	Annex I.B.2: Organization and Personnel
4.4.2 Training, Awareness, and Competence of personnel shall be established and maintained to ensure policy and procedure implementation	4.18 Training of personnel on activities affecting quality shall be established, maintained, and documented.	4.3.4 Communication and training on environmental policy and procedures of personnel; training records shall be maintained.	Annex I.B.2: Communication and training on policy and procedures of personnel; records shall be maintained of all communications.
4.4.3 Communication regarding EMS policy, both internal and external, shall be maintained and documented.	4.1.2.3 Establishes a liaison with external parties by management		
4.4.6 Operations control shall be identified and documented; must communicate to contractors and suppliers as they have potential to impact environmental objectives and policy.	4.2.2 Quality system procedures 4.6 Purchasing 4.3 Contract review 4.4 Design control 4.7 Control of customer supplied product 4.8 Product identification and traceability 4.9 Process control 4.15 Handling, storage, packaging, preservation, and delivery 4.19 Servicing	4.8.1, 2 Controls to ensure activities, procedures, and work instructions dealing with procurement and suppliers shall be monitored and coordinated. 4.3.5 Contractors shall be aware of requirements.	Annex I.B.4: Controls shall establish operating procedures affecting environment; includes documentation, monitoring, and procedures that deal with procurement and contractors.

ISO 14001	ISO 9001	BS 7750	EMAS
4.4.7 A procedure to implement and maintain emergency preparedness and response activities for the prevention of accidents and emergency situations shall be implemented and tested.	Not applicable	4.4.2 The environmental effects register shall include procedures for managing accidents and emergency situations	Annex I.C.9: Prevention and limitation of environmental accidents shall be addressed. Annex I.C.10: Contingency procedures in cases of environmental accidents shall be addressed.
4.5.1 Monitoring and measurement procedures shall be established to verify conformance and performance against objectives and targets and policy.	4.10 Inspection and testing 4.11 Control of inspection, measuring, and test equipment. 4.12 Inspection and test status 4.20 Statistical techniques	4.8.3 Compliance to verify, measure and test compliance with specified requirements shall be documented and maintained to assess activities	Annex I.B.4: Procedures for monitoring shall be implemented to ensure policy requirements are met, documented, and appropriate activities are assessed.
4.5.2 Procedures for Nonconformance and corrective and preventive action shall be established; define responsibility and handle nonconformance; changes from mitigating action must be recorded.	4.13 Control of nonconforming product to prevent unintended use 4.14 Corrective and preventive action must be taken to ensure quality; changes must be recorded.	4.8.4 Documented procedure for non-compliance and corrective action shall be defined, established and maintained; controls must be applied and changes to procedures recorded.	Annex I.B.4: Noncompliance and corrective action: Noncompliances shall be identified, evaluated, corrected, and controlled and any changes recorded.
4.5.4 Periodic EMS audit, based on relevancy of activities shall be conducted to determine conformity, implementation, and maintenance of standards; results reported to management.	4.17 Periodic internal audits shall be performed to verify whether activities and results comply with quality standards; must be recorded and reviewed.	4.10 Environmental management audits shall be maintained to assess conformity, effectiveness of EMS 4.10.1 General 4.10.2 Audit program 4.10.3 Audit protocols and procedures	Annex I.B.6: Audits shall be done periodically to verify conformance and the effectiveness of the implementation and adherence to the EMS policy.
4.6 Periodic management reviews shall be done to ensure effectiveness and adequacy; they shall address possible need for changes and be documented	4.1.3 Management review shall be performed at periodic intervals to ensure effectiveness of policy; records shall be maintained of the reviews.	4.11 Periodic reviews of EMS to satisfy standards shall be conducted to ensure effectiveness and address need for changes; reviews shall be documented.	Annex I.D: Regular checks of Good Mgmnt Practices based on "principles of action" in policy to ensure continual improvement. Annex I.B.1: Periodic reviews of environmental policy and objectives.

Appendix D
International Chamber
of Commerce Charter
for Sustainable Development

1. Corporate Policy:
To recognize environmental management as among the highest corporate priorities and as a key determinant to sustainable development; to establish policies, programs, and practices for conducting operations in an environmentally sound manner.

2. Integrated Management:
To integrate these policies, programs, and practices fully into each business as an essential element of management in all its functions.

3. Process of Improvement:
To continue to improve policies, programs, and environmental performance, taking into account technical developments, scientific understanding, consumer needs and community expectations, with legal regulations as starting point; and to apply the same environmental criteria internationally.

4. Employee Education:
To educate, train, and motivate employees to conduct their activities in an environmentally responsible manner.

5. Prior Assessment:
To assess environmental impacts before starting a new activity or project and before decommissioning a facility or leaving a site.

6. Products and Services:
To develop and provide products or services that have no undue environmental impact and are safe in their intended use, that are efficient in their consumption of energy and natural resources, and that they can be recycled, reused, or disposed of safely.

7. Customer Advice:
To advise, and where relevant, educate customers, distributors, and the public in the safe use, transportation, storage, and disposal of products provided; and to apply similar considerations to the provisions of services.

8. Facilities and Operations:
To develop, design, and operate facilities and conduct activities taking into consideration the efficient use of energy and materials, the sustainable use of renewable resources, the minimization of adverse environmental impact and waste generation, and the safe and responsible disposal of residual wastes.

9. Research:
To conduct or support research on the environmental impacts of raw materials, products, processes, emissions, and wastes associated with the enterprise and on the means of minimizing such adverse impacts.

10. Precautionary Approach:

To modify the manufacture, marketing, or use of products or services or the conduct of activities, consistent with scientific and technical understanding, to prevent serious or irreversible environmental degradation.

11. Contractors and Suppliers:

To promote the adoption of these principles by contractors acting on behalf of the enterprise, encouraging and, where appropriate, requiring improvements in their practices to make them consistent with those of the enterprise; and to encourage the wider adoption of these principles by suppliers.

12. Emergency Preparedness:

To develop and maintain, where significant hazards exist, emergency preparedness plans in conjunction with the emergency services, relevant authorities and the local community, recognizing potential transboundary impacts.

13. Transfer of Technology:

To contribute to the transfer of environmentally sound technology and management methods throughout the industrial and public sectors.

14. Contributing to the Common Effect:

To contribute to the development of public policy and to business, governmental and intergovernmental programs and educational initiatives that will enhance environmental awareness and protection.

15. Openness to Concerns:

To foster openness and dialogue with employees and the public, anticipating and responding to their concerns about potential hazards and impacts of operations, products, wastes or services, including those of transboundary or global significance.

16. Compliance and Reporting:

To measure environmental performance; to conduct regular environmental audits and assessments of compliance with company requirements, legal requirements and these principles; and periodically to provide appropriate information to the Board of Directors, shareholders, employees, the authorities, and the public.

Appendix E
Rio Declaration on Environment and Development

The United Nations Conference on Environment and Development, having met at Rio de Janeiro June 3–14, 1992, reaffirming the Declaration of the United Nations Conference on the Human Environment, adopted at Stockholm on June 16, 1972, and seeking to build upon it, with the goal of establishing a new and equitable global partnership through the creation of new levels of cooperation among States, key sectors of societies and people, working towards international agreements which respect the interests of all and protect the integrity of the global environmental and developmental system, recognizing the integral and interdependent nature of the Earth, our home proclaims that:

Principle 1   Human beings are at the centre of concerns for sustainable development. They are entitled to a healthy and productive life in harmony with nature.

Principle 2   States have, in accordance with the Charter of the United Nations and the principles of international law, the sovereign right to exploit their own resources pursuant to their own environmental and developmental policies, and the responsibility to ensure that activities within their jurisdiction or control do not cause damage to the environment of other States or of areas beyond the limits of national jurisdiction.

Principle 3   The right to development must be fulfilled so as to equitably meet developmental and environmental needs of present and future generations.

Principle 4   In order to achieve sustainable development, environmental protection shall constitute an integral part of the development process and cannot be considered in isolation from it.

Principle 5   All States and all people shall cooperate in the essential task of eradicating poverty as an indispensable requirement for sustainable development, in order to decrease the disparities in standards of living and better meet the needs of the majority of the people of the world.

Principle 6   The special situation and needs of developing countries, particularly the least developed and those most environmentally vulnerable, shall be given special priority. International actions in the field of environment and development should also address the interests and needs of all countries.

Principle 7   States shall cooperate in a spirit of global partnership to conserve, protect and restore the health and integrity of the Earth's eco-system. In view of the different contributions to global environmental degradation, States have common but differentiated responsibilities. The developed countries acknowledge the responsibility that they bear in the international pursuit of sustainable development in view of the pressures their societies place on the global environment and of the technologies and financial resources they command.

Principle 8   To achieve sustainable development and a higher quality of life for all people, States should reduce and eliminate unsustainable patterns of production and consumption and promote appropriate demographic policies.

Principle 9   States should cooperate to strengthen endigenous capacity-building for sustainable development by improving scientific understanding through exchanges of scientific and technological knowledge, and by enhancing the development, adaptation, diffusion, and transfer of technologies, including new and innovative technologies.

Principle 10   Environmental issues are best handled with the participation of all concerned citizens, at the relevant level. At the national level, each individual shall have appropriate access to information concerning the environment that is held by public authorities, including information on hazardous materials and activities in their communities, and the opportunity to participate in decision-making processes. States shall facilitate and encourage public awareness and participation by making information widely available. Effective access to judicial and administrative proceedings, including redress and remedy, shall be provided.

Principle 11   States shall enact effective environmental legislation. Environmental standards, management objectives and priorities should reflect the environmental and developmental context to which they apply. Standards applied by some countries can be inappropriate and of unwarranted economic and social cost to other countries, in particular to developing countries.

Principle 12   States should cooperate to promote a supportive and open international economic system that would lead to economic growth and sustainable development in all countries, to better address the problems of environmental degradation. Trade policy measures for environmental purposes should not constitute a means of arbitrary or unjustifiable discrimination or a disguised restriction on international trade. Unilateral actions to deal with environmental challenges outside the jurisdiction of the importing country should be avoided. Environmental measures addressing transboundary or global environmental problems should, as far as possible, be based on an international consensus.

Principle 13   States shall develop national law regarding liability and compensation for the victims of pollution and other environmental damage. State shall also cooperate in an expeditious and more determined manner to develop further international law regarding liability and compensation for adverse effects of environmental damage caused by activities within their jurisdiction or control to areas beyond their jurisdiction.

Principle 14 States should effectively cooperate to discourage or prevent the relocation and transfer to other States of any activities and substances that cause severe environmental degradation or are found to be harmful to human health.

Principle 15 In order to protect the environment, the precautionary approach shall be widely applied by States according to their capabilities. Where there are threats of serious or irreversible damage, lack of full scientific certainty shall not be used as a reason for postponing cost-effective measures to prevent environmental degradation.

Principle 16 National authorities should endeavour to promote the internalisation of environmental costs and the use of economic instruments, taking into account the approach that the polluter should, in principle, bear the cost of pollution, with due regard to the public interest and without distorting international trade and investment.

Principle 17 Environmental impact assessment, as a national instrument, shall be undertaken for proposed activities that are likely to have a significant adverse impact on the environment and are subject to a decision of a component national authority.

Principle 18 States shall immediately notify other States of any natural disasters or other emergencies that are likely to produce sudden harmful effects on the environment of those States. Every effort shall be made by the international community to help States so afflicted.

Principle 19 States shall provide prior and timely notification and relevant information to potentially affected States on activities that can have a significant adverse transboundary environmental effect and shall consult with those States at an early stage and in good faith.

Principle 20 Women have a vital role in environmental management and development. Their full participation is therefore essential to achieve sustainable development.

Principle 21 The creativity, ideals and courage of the youth of the world should be mobilised to forge a global partnership in order to achieve sustainable development and ensure a better future for all.

Principle 22 Indigenous people and their communities, and other local communities, have a vital role in environmental management and development because of their knowledge and traditional practices. States should recognise and duly support their identity, culture and interest and enable their effective participation in the achievement of sustainable development.

Principle 23 The environment and natural resources of people under oppression, domination and occupation shall be protected.

Principle 24 Warfare is inherently destructive of sustainable development. States shall therefore respect international law providing protection for the environment in times of armed conflict and cooperate in its further development, as necessary.

Principle 25 Peace, development and environmental protection are interdependent and indivisible.

Principle 26 States shall resolve all their environmental disputes peacefully and by appropriate means in accordance with the Charter of the United Nations.

Principle 27 States and people shall cooperate in good faith and in a spirit of partnership in the fulfillment of the principles embodied in this Declaration and in the further development of international law in the field of sustainable development.

Appendix F
ISO 14001
Self-Assessment Checklist

ELEMENT 4.2 — ENVIRONMENTAL POLICY

Requirement *"Top management shall define the organization's environmental policy and ensure that it ...*

	Yes	No
(a) is appropriate to ... of its activities, products or services;		
(i) *nature*	[ ]	[ ]
(ii) *scale*	[ ]	[ ]
(iii) *environmental impacts* (see 4.3.1)	[ ]	[ ]
(b) includes a commitment to...		
(i) *continual improvement*	[ ]	[ ]
(c) includes a commitment to...		
(i) *comply with relevant environmental legislation and regulations*	[ ]	[ ]
(ii) *other requirements to which the organization subscribes*	[ ]	[ ]
(d) provides the framework for...		
(i) *setting and reviewing environmental objectives and targets*	[ ]	[ ]
(e) is...		
(i) *documented*	[ ]	[ ]
(ii) *implemented*	[ ]	[ ]
(iii) *maintained*	[ ]	[ ]
(iv) *communicated to all employees*	[ ]	[ ]

How?

_____

_____

_____

	Yes	No
(f) is ...		
(i) *available to the public*	[ ]	[ ]

How?

_____

_____

_____

ELEMENT 4.3.1 — ENVIRONMENTAL ASPECTS

Requirement *"The organization shall establish and maintain (a) procedure(s) to identify the environmental aspects of its activities, products, or services...*

                                                                        Yes   No
(a) it can...
    (i) *control*   [ ]   [ ]

What are they?

_____

_____

(b) over which it can be expected to have...
    (i) *an influence*   [ ]   [ ]

What are they?

_____

_____

(c) in order to determine those which...
    (i) *can or can have significant impacts on the environment.*   [ ]   [ ]

How does the organization determine what is "significant"?

_____

_____

Requirement *"The organization shall ensure that...*

(d) the aspects related to these significant aspects are...
    (i) *considered in setting its environmental objectives*   [ ]   [ ]

Evidence?

_____

_____

(e) this information is...
    (i) *kept up-to-date*   [ ]   [ ]

Evidence?

_____

_____

ELEMENT 4.3.2 — LEGAL AND OTHER REQUIREMENTS

Requirement   *"The organization shall establish and maintain (a) procedure(s) to*

	Yes	No
(a) identify, and have access to...		
(i) *legal and*	[ ]	[ ]
(ii) *other requirements*	[ ]	[ ]

How are requirements assessed?

_____

_____

(b) to which the organization subscribes, that are...
  (i) *applicable to the environmental aspects of its activities
      products, or services*   [ ]   [ ]

How does the organization determine what is "applicable"?

_____

_____

ELEMENT 4.3.3 — OBJECTIVES AND TARGETS

Requirement   *"The organization shall establish and maintain documented*

	Yes	No
(a) environmental objectives and targets at each...		
(i) *relevant function and*	[ ]	[ ]
(ii) *level within the organization*	[ ]	[ ]

What are the relevant functions and levels?

_____

_____

Requirement   *"When establishing and reviewing its objectives, an organization shall consider*

	Yes	No
(a) *the legal and other requirements*	[ ]	[ ]
(b) *its significant environmental aspects*	[ ]	[ ]
(c) *its technological options*	[ ]	[ ]
(d) *its financial, operational, and business requirements, and*	[ ]	[ ]
(e) *the views of interested parties.*	[ ]	[ ]

Evidence:

_____

_____

Requirement   *"The objectives and targets shall be...*

	Yes	No
(a) *consistent w/the environmental policy, including*	[ ]	[ ]
(b) *the commitment to prevention of pollution*	[ ]	[ ]

Evidence:

_____

_____

ELEMENT 4.3.4 — ENVIRONMENTAL MANAGEMENT PROGRAMS

Requirement   *"The organization shall establish and maintain (a) program(s) for achieving its objectives and targets.*

	Yes	No
(a) It shall include...		
(i) *designation of responsibility*	[ ]	[ ]
(ii) *the means*	[ ]	[ ]
(iii) *the timeframe*	[ ]	[ ]

by which they are to be achieved.

Where is responsibility designated?

_____

_____

How is the means and timeframe established?

_____

_____

Requirement   *"If a project relates to new developments and new or modified activities, products or services...*

(a) program(s) shall be...

	Yes	No
(i) amended where relevant to ensure that environmental management applies to such projects	[ ]	[ ]

Is there a procedure to deal with this?

_____

_____

ELEMENT 4.4.1 — STRUCTURE AND RESPONSIBILITY

Requirement   *"Roles, responsibilities, and authorities shall be...*

	Yes	No
(a) *defined*	[ ]	[ ]
(b) *documented*	[ ]	[ ]
(c) *communicated*	[ ]	[ ]

Evidence:

_____

_____

Requirement   *"Management shall provide...*

(a) *resources essential to the implementation and control of the EMS*   [ ]   [ ]

What type of resources (HR, special skills, technology, financial, etc.)

_____

_____

Requirement   *"The organization's top management shall appoint (a) specific management representative(s) who, irrespective of other responsibilities, shall have defined roles, responsibilities, and authority for...*

(a) ensuring that EMS requirements are...

(i) *established*	[ ]	[ ]
(ii) *implemented*	[ ]	[ ]
(iii) *maintained*	[ ]	[ ]

Where is this defined?

_____

_____

(b) *reporting* on the performance of the EMS to top management
for review and as a basis for improvement of the EMS   [ ]   [ ]

ELEMENT 4.4.2 — TRAINING, AWARENESS, AND COMPETENCE

Requirement *"The organization shall identify training needs."*

	Yes	No
Procedure in place?	[ ]	[ ]

Requirement *"It shall require that all personnel whose work may create a significant impact upon the environment, have received appropriate training."*

Evidence:

_____

_____

Requirement *"It shall establish and maintain procedures to make its employees or members at each relevant function and level aware of...*

(a) the importance of conformance with the...

	Yes	No
(i) *environmental policy*	[ ]	[ ]
(ii) *procedures*	[ ]	[ ]
(iii) *the requirements of the EMS*	[ ]	[ ]

Procedure available?

_____

_____

(b) the significant environmental impacts...

	Yes	No
(i) *actual or potential of their work activities*	[ ]	[ ]
(ii) *the environmental benefits of improved personal performance*	[ ]	[ ]

(c) their roles and responsibilities in...

	Yes	No
(i) *achieving conformance with the environmental policy and procedures*	[ ]	[ ]
(ii) *and with the requirements of the EMS including emergency preparedness and response requirements*	[ ]	[ ]

Requirement *"Personnel performing the tasks which can cause significant environmental impacts shall be competent on the basis of appropriate education, training, and/or experience."*

Evidence:

_____

_____

ELEMENT 4.4.3 — COMMUNICATION

Requirement   *"With regards to its environmental aspects and EMS the organiza-tion shall establish and maintain procedures for...*

	Yes	No
(a) *internal communication* between the various levels and functions of the organization	[ ]	[ ]

Procedure?

_____

_____

(b) *recording, documenting*, and *responding* to relevant communication from external interested parties   [ ]   [ ]

Procedure?

_____

_____

Requirement   *"The organization shall...*

(a) *consider* processes for external communication on its significant environmental aspects and   [ ]   [ ]
(b) *record* its decision   [ ]   [ ]

Evidence:

_____

_____

ELEMENT 4.4.4 — ENVIRONMENTAL MANAGEMENT SYSTEM DOCUMENTATION

Requirement   *"The organization shall establish and maintain information, in paper or electronic form, to*

	Yes	No
(a) *describe the core elements* of the management system and their interaction	[ ]	[ ]
(b) *provide direction to related documentation*	[ ]	[ ]

What is the documented evidence?

_____

_____

ELEMENT 4.4.5 — DOCUMENT CONTROL

Requirement   *"The organization shall establish and maintain procedures for controlling all documents required by this International Standard to ensure that*

	Yes	No
(a) *they can be located*	[ ]	[ ]
(b) they are periodically *reviewed*, revised as necessary, and *approved* for adequacy by authorized personnel	[ ]	[ ]
(c) the *current versions* of relevant documents are *available* at all locations where operations essential to the effective functioning of the environmental management system are performed	[ ]	[ ]
(d) *obsolete documents are promptly removed* from all points of issue, or otherwise assured against unintended use	[ ]	[ ]
(e) any *obsolete documents retained for legal* and/or knowledge preservation purposes are suitably identified	[ ]	[ ]

Procedure(s) ?

_____

_____

Requirement   *"Documentation shall be ...*

	Yes	No
(a) *legible*	[ ]	[ ]
(b) *dated* (with dates of revision)	[ ]	[ ]
(c) readily *identifiable*	[ ]	[ ]
(d) *maintained* in an orderly manner	[ ]	[ ]
(e) *retained* for a specified period	[ ]	[ ]

Requirement   *"Procedures and responsibilities shall be established and maintained concerning the...*

	Yes	No
(a) *creation* and	[ ]	[ ]
(b) *modification* of the various types of documents	[ ]	[ ]

Procedures?

_____

_____

ELEMENT 4.4.6 — OPERATIONAL CONTROL

Requirement   *"The organization shall identify those operations and activities that are associated with the identifiable significant environmental aspects in line with its policy, objectives, and targets. The organization shall plan these activities,* including maintenance, *in order to ensure that they are carried out under specified conditions by*

	Yes	No
(a) establishing and maintaining documented procedures to cover situations where their *absence could lead to deviations* from the environmental policy and the objectives and targets	[ ]	[ ]

Evidence:

_____   _____

_____

(b) stipulating *operating criteria* in the procedures	[ ]	[ ]

Evidence:

_____

_____

(c) establishing and maintaining procedures related to the identifiable significant environmental aspects of

(i) *goods and services* used by the organization and	[ ]	[ ]
(ii) *communicating* relevant procedures and requirements *to suppliers and contractors*	[ ]	[ ]

Procedures?

_____

_____

ELEMENT 4.4.7 — EMERGENCY PREPAREDNESS AND RESPONSE

Requirement   *"The organization shall establish and maintain procedures to*

	Yes	No
(a) *identify potential for and respond* to accidents and emergency situations	[ ]	[ ]
(b) for *preventing and mitigating* the environmental impacts that may be associated with them	[ ]	[ ]

Requirement   *"The organization shall ...*

(a) *review and*	[ ]	[ ]
(b) *revise*	[ ]	[ ]

where necessary, its emergency preparedness and response procedures, in particular, after the occurrence of accidents or emergency situations.

Evidence:

_____

_____

Requirement   *"The organization shall establish and maintain...*

(a) a documented *procedure for periodically evaluating compliance* with relevant environmental legislation and regulations	[ ]	[ ]

Procedure?

_____

_____

ELEMENT 4.5.1 — MONITORING AND MEASUREMENT

Requirement *"The organization shall establish and maintain procedures to*

	Yes	No
(a) *monitor and measure*, on a regular basis, the *key* characteristics of its operations and activities that can have a significant impact on the environment.	[ ]	[ ]

This shall include...

	Yes	No
(a) the *recording of information* to track performance	[ ]	[ ]
(b) *relevant operational controls*	[ ]	[ ]
(c) conformance with the organization's environmental objectives and targets	[ ]	[ ]

Procedures?

_____

_____

Requirement *"Monitoring equipment shall be...*

	Yes	No
(a) *calibrated*	[ ]	[ ]
(b) *maintained*	[ ]	[ ]
(c) *records* of this process shall be retained according to the organization's procedures	[ ]	[ ]

Evidence:

_____

_____

Requirement *"The organization shall establish and maintain a documented procedure for*

	Yes	No
(a) *periodically evaluating compliance* with relevant environmental legislation and regulations	[ ]	[ ]

Procedure?

_____

_____

ELEMENT 4.5.2 — NONCONFORMANCE AND CORRECTIVE AND PREVENTIVE ACTION

Requirement   *"The organization shall establish and maintain procedures for*

	Yes	No
(a) defining responsibility and authority for...		
(i) *handling and investigating* nonconformance	[ ]	[ ]
(ii) taking *action* to mitigate any impacts caused	[ ]	[ ]
(iii) for *initiating and completing* corrective and preventive action	[ ]	[ ]

Procedure?

_____

_____

Requirement   *"Any corrective or preventive action taken to eliminate the causes of actual or potential nonconformances shall be...*

	Yes	No
(a) *appropriate* to the magnitude of problems	[ ]	[ ]
(b) *commensurate* with the environmental impact encountered	[ ]	[ ]

Evidence:

_____

_____

Requirement   *The organization shall*

	Yes	No
(a) *implement, and*	[ ]	[ ]
(b) *record*	[ ]	[ ]

any changes in the documented procedures resulting from corrective and preventive action

Evidence:

_____

_____

ELEMENT 4.5.3 — RECORDS

Requirement *"The organization shall establish and maintain procedures for*

	Yes	No
(a) the *identification*,	[ ]	[ ]
(b) *maintenance*, and	[ ]	[ ]
(c) *disposition* of environmental records	[ ]	[ ]

Procedures?

_____

_____

Requirement *"These records shall include*

(a) *training records*,	[ ]	[ ]
(b) *the results of audits, and*	[ ]	[ ]
(c) *reviews*	[ ]	[ ]

Evidence:

_____

_____

Requirement *"Environmental records shall be...*

(a) *legible, identifiable, and traceable*　　　　　　[ ]　[ ]

to the activity, product or service involved. Environmental records shall be...

(b) *stored and maintained*　　　　　　　　　　　[ ]　[ ]

in such a way that they are readily retrievable and protected against damage, deterioration, or loss.

(c) Their *retention times* shall be established and recorded　[ ]　[ ]

Requirement *"Records shall be...*

(a) *maintained*, as appropriate to the system and to the organization, to demonstrate conformance to the requirements of this International Standard　　　　　　　　　　　　　　　[ ]　[ ]

ELEMENT 4.5.4 — ENVIRONMENTAL MANAGEMENT SYSTEM AUDIT

Requirement   *"The organization shall establish and maintain (a) program(s)and procedures for periodic environmental management system audits to be carried out, in order to...*

	Yes	No
(a) determine whether or not the environmental management system...		
(i) *conforms to planned arrangements* for environmental management including the requirements of this International Standard	[ ]	[ ]
(ii) has been properly *implemented and maintained*	[ ]	[ ]
(b) provide *information* on the results of audits *to management*	[ ]	[ ]

Evidence:

_____

_____

Requirement   *"The organization's audit program, including any schedule, shall be based on...*

(a) the environmental *importance* of the activity concerned	[ ]	[ ]
(b) the results of *previous audits*	[ ]	[ ]

Criteria?

_____

_____

Requirement   *"In order to be comprehensive, the audit procedures shall cover...*

(a) the audit *scope*	[ ]	[ ]
(b) *frequency*	[ ]	[ ]
(c) *methodology*	[ ]	[ ]
(d) *responsibilities*	[ ]	[ ]
(e) *requirements* for conducting audits	[ ]	[ ]
(f) *reporting* results	[ ]	[ ]

Procedure?

_____

_____

ELEMENT 4.6 — MANAGEMENT REVIEW

Requirement   *"The organization's top management shall, at intervals that it determines...*

	Yes	No
(a) review the environmental management system, to ensure...		
(i) *its continuing suitability*	[ ]	[ ]
(ii) *adequacy*	[ ]	[ ]
(iii) *effectiveness*	[ ]	[ ]
(b) The management review process shall...		
(i) ensure that the necessary *information is collected* to allow management to carry out this evaluation.	[ ]	[ ]
(ii) *be documented*	[ ]	[ ]

Evidence:

_____

_____

Requirement   *"The management review shall...*

	Yes	No
(a) address the possible need for changes to		
(i) *policy*	[ ]	[ ]
(ii) *objectives*	[ ]	[ ]

Evidence:

_____

_____

	Yes	No
(b) in the light of ...		
(i) EMS *audit results*	[ ]	[ ]
(ii) *changing circumstances*	[ ]	[ ]
(iii) the commitment of *continual improvement*	[ ]	[ ]

Evidence:

_____

_____

Appendix G
Keidanren (Japan Federation of Economic Organizations) Global Environment Charter

BASIC PHILOSOPHY

A company's existence is closely bound up with the global environment as well as with the community it is based in. In carrying on its activities, each company must have respect for human dignity and strive toward a future society where the global environment is protected.

We must aim to construct a society whose members cooperate together on environmental problems, a society where sustainable development on a global scale is possible, where companies enjoy a relationship of mutual trust with local citizens and consumers, and where they vigorously and freely develop their operations while preserving the environment. Each company must aim at being a good global corporate citizen, recognizing that grappling with environmental problems is essential to its own existence and its activities.

GUIDELINES FOR CORPORATE ACTION

Companies must carry on their business activities to contribute to the establishment of a new economic social system for realizing an environmentally protective society leading to the sustainable development.

1. GENERAL MANAGEMENT PRINCIPLES

Companies should always carry on their business activities to contribute to the establishment of a new economic social system for realizing an environmentally protective society leading to the sustainable development.

2. CORPORATE ORGANIZATION

(a) Companies shall establish an internal system to handle environmental issues by appointing an executive and creating an organization in charge of environmental problems.

(b) Environmental regulations shall be established for company activities and these shall be observed. Such internal regulations shall include goals for reducing the load on the environment. An internal inspection to determine how well the environmental regulations are being adhered to shall be carried out at least once a year.

3. CONCERN FOR THE ENVIRONMENT

(a) All company activities, beginning with siting of production facilities, shall be scientifically evaluated for their impact on the environment and necessary counter-measures shall be implemented.

(b) Care shall be taken in the research, design, and development stages of making a product to lessen the possible burden on the environment at each level of its production, distribution, appropriate use, and disposal.

(c) Companies shall strictly observe all national and local laws and regulations for environmental protection and, where necessary, they shall set additional standards of their own.

(d) When procuring materials, including materials for production, companies shall endeavor to purchase those that are superior for conserving resources, preserving the environment, and enhancing recycling.

(e) Companies shall employ technologies that allow efficient use of energy and preservation of the environment in their production and other activities. Companies shall endeavor to use resources efficiently and reduce waste products through recycling, and shall appropriately deal with pollutants and waste products.

4. TECHNOLOGY DEVELOPMENT

(a) In order to help solve global environmental problems, companies shall endeavor to develop and supply innovative technologies, products and services that allow conservation of energy, and other resources together with preservation of the environment.

5. TECHNOLOGY TRANSFERS

(a) Companies shall seek appropriate means for the domestic and overseas transfer of their technologies, know-how, and expertise for dealing with environmental problems and conserving energy and other resources.

(b) In participating in official development assistance projects, companies shall carefully consider environmental and anti-pollution measures.

6. EMERGENCY MEASURES

(a) If environmental problems ever occur as a result of an accident in the course of company activities or deficiency in a product, companies shall adequately explain the situation to all concerned parties and take appropriate measures, using their technologies and human and other resources, to minimize the impact on the environment.

(b) Even when a major disaster or environmental accident occurs outside of a company's responsibility, it shall still actively provide technological and other appropriate assistance.

7. Public Relations and Education

(a) Companies shall actively publicize information and carry out educational activities concerning their measures for protecting the environment, maintaining the ecosystems, and ensuring health and safety in their activities.

(b) The employees shall be educated to understand the importance of daily close management to ensure the prevention of pollution and the conservation of energy and other resources.

(c) Companies shall provide users with information of the appropriate use and disposal, including recycling, of their products.

8. Community Relations

(a) As community members, companies shall actively participate in activities to preserve the community environment and support employees who engage in such activities on their own initiative.

(b) Companies shall promote dialogue with people in all segments of society over operational issues and problems seeking to achieve mutual understanding and strengthen cooperative relations.

9. Overseas Operations

(a) Companies developing operations overseas shall observe the Ten-Points-Environmental Guidelines for the Japanese Enterprises Operating Abroad in Keidanren's Basic Views of the Global Environmental Problems.

10. Contribution to Public Policies

(a) Companies shall work to provide information gained from their experiences to administrative authorities, international organizations, and other bodies formulating environmental policy, as well as participate in dialogue with such bodies, in order that more rational and effective policies can be formulated.

(b) Companies shall draw on their experiences to propose rational systems to administrative authorities and international organizations concerning formulation of environmental policies to offer sensible advice to consumers on lifestyles.

11. Response to Global Problems

(a) Companies shall cooperate in scientific research on the causes and effects of such problems as global warming and they shall also cooperate in the economic analysis of possible counter-measures.

(b) Companies shall actively work to implement effective and rational measures to conserve energy and other resources even when such environmental problems have not been fully elucidated by science.

(c) Companies shall play an active role when the private sector's help is sought to implement international environmental measures, including work to solve the problems of poverty and over-population in developing countries.

Environmental, Health and Safety (EHS) Questionnaire for Existing and Proposed Critical Materials and/or Components Suppliers

Contents

I. EHS Management Overview and Background for the Facility Supplying Raychem (7 questions; 200 points)

II. EHS Systems and Procedures at the Facility Supplying Raychem (7 questions; 450 points)

III. Internal Audits of the Facility Supplying Raychem (2 questions; 120 points)

IV. EHS Performance and Continuous Improvement at the Facility Supplying Raychem (4 questions; 230 points)

Supplier Name: _____ Date: _____

Supplier Address: _____

Address of Specific Facility Supplying Products for Joint Venture:

_____

Name of Person Completing Form: _____

Title: _____

Please answer the following questions for the specific facility responsible for supplying Raychem. Check appropriate boxes. Attach additional comments or information as necessary. Please write N/A (Not Applicable) in the margin when the subject will never apply at this location. Write a question mark (?) in the margin if you do not know the answer.

I. EHS Management Overview & Background for the Facility Supplying Raychem	Yes	No	Pts.*
1. Does the facility have a written EHS Policy Statement? (XO-20) If yes, does the policy specifically call for:	❑	❑	____
a. compliance with EHS regulations? (XO-10)	❑	❑	____
b. continuous improvement? (XO-10)	❑	❑	____
c. product stewardship? (XO-10)	❑	❑	____
Please attach a copy.			
2a. Is a management representative assigned responsibility for facilitating compliance with EHS regulations? (XO-20) If yes, please give name, title and phone no.:_____	❑	❑	____
b. Has management established a site EHS committee? (XO-30) If yes, please attach a copy of its charter or mission statement.	❑	❑	____
3. Have long term EHS goals been established for the facility? (XO-20)	❑	❑	____
4. Are quantitative annual EHS objectives established at various levels of your organization at the facility (e.g., reduction targets for waste disposal, air emissions, water discharges, accidents or incidents)? (XO-30)	❑	❑	____
5. Does the site manager (or designee) maintain a written list or registry of EHS regulatory requirements that apply to the operations of the facility? (XO-10) If yes, does it include requirements related to:	❑	❑	____
a. worker protection? (XO-10)	❑	❑	____
b. environmental and community protection? (XO-10)	❑	❑	____
c. property protection? (XO-10)	❑	❑	____
d. product stewardship? (XO-10)	❑	❑	____
6. Please estimate the number (a range is acceptable) of temporary employees (____) and contractors' employees (____) typically present on site.			
7a. Does any other company or corporation (e.g., contract suppliers to you) occupy the same site or any part of it? If yes, please identify them and describe.	❑	❑	
b. Are there adjacent operations (public or private) that are generally considered by the local community, insurers, or experienced EHS professionals to represent substantial risks to people, property, or the environment (e.g., oil refineries, chemical processing, hazardous waste processing facilities)? If yes, please identify and describe them on an attached page.	❑	❑	
Total for Section I			____

II. EHS Systems and Procedures at the Facility Supplying Raychem	Yes	No	Pts.*
8. Is there a site EHS manual of guidelines and procedures? (XO-30) If yes, please supply a copy of the table of contents.	❑	❑	____
9. Does the facility have systems or activities with written procedures in the following areas for:			
a. Worker Protection:			

i. Site Work Permit system for employees & contractors? If yes, check all that apply:
   ❑ Confined Space Entry Permit (XO-5)
   ❑ Excavation Permit (XO-5)
   ❑ Fork Lift Operating Permit (XO-5)
   ❑ Hot Work Permit (XO-5)
   ❑ Fire Protection System Impairment Permit (XO-5)
   ❑ Utility Interruption Permit (XO-5)
	❑	❑	____

ii. Employee drug testing? (XO-10)	❑	❑	____
iii. Accident investigation, reporting, and follow-up? (XO-10)	❑	❑	____
iv. Near-miss investigation, reporting, and follow-up? (XO-10)	❑	❑	____
b. Property Protection:			
i. Routine testing and maintenance of fire detection/alarms and suppression systems? (XO-10)	❑	❑	____
ii. Annual testing of fire water pumps? (XO-10)	❑	❑	____
c. Environmental Protection:			
i. Process air emissions tracking or monitoring? (XO-10)	❑	❑	____
ii. Process wastewater discharge tracking or monitoring? (XO-10)	❑	❑	____
iii. Ensuring solid and liquid (containerized) wastes from manufacturing are identified, accumulated, stored, treated, and properly disposed of in accordance with applicable local regulations? (XO-10)	❑	❑	____
iv. Storm water pollution prevention? (XO-10)	❑	❑	____
v. Ground water pollution prevention? (XO-10)	❑	❑	____
vi. Fire-fighting water containments? (XO-10)	❑	❑	____
vii. Energy conservation? (XO-10)	❑	❑	____
d. Product Stewardship:			
i. Recycling or reuse of products? (XO-10)	❑	❑	____
ii. Minimization of product packaging materials? (XO-10)	❑	❑	____
10a. Does the facility have a written emergency response plan? (XO-20) If yes, please supply a copy of the table of contents.	❑	❑	____
b. How frequently are emergency response drills conducted? ❑ Quarterly (20) ❑ Semiannually (10) ❑ Annually (5)			____

11a. Does the facility have a written Business Recovery Plan for the facility? (XO-30) If yes, please supply a copy of the table of contents and proceed to 11b and 11c. If no, go to question 12. ❏ ❏ ____

  b. In preparing the Business Recovery Plan, did you provide recovery procedures for:

   1. accidents at your facility (e.g., fires, chemical spills, equipment failures) (XO-10) ❏ ❏ ____

   2. natural disasters at your facility (e.g., wind storm, flood, lightning strike, earthquake) (XO-10) ❏ ❏ ____

   3. accidents at an adjacent facility that could adversely affect your facility (XO-10) ❏ ❏ ____

   4. failure of your utility suppliers to deliver for an extended period (XO-10) ❏ ❏ ____

   5. failure of a material or component supplier to deliver for an extended period (internal or external to your company) (XO-10) ❏ ❏ ____

  c. Is the plan reviewed and tested on an annual basis? (XO-10) ❏ ❏ ____

12a. Is EHS training (including hazard communication) provided to all employees upon hire as well as on a routine basis? (XO-20) ❏ ❏ ____

  b. If temporary employees and/or contractors' employees are routinely working on your site, do they receive:

   1. a site EHS orientation before beginning work that includes site rules? (XO-10) ❏ ❏ ____

   2. briefings on area-specific EHS hazards and area-specific EHS controls where they work? (XO-10) ❏ ❏ ____

13. Does the facility specify, procure, or use recycled materials in:

  a. products? (XO-10) ❏ ❏ ____

  b. packaging? (XO-10) ❏ ❏ ____

  c. office paper and supplies? (XO-10) ❏ ❏ ____

14. For chemical suppliers:

  a. Is the facility a participant in the Responsible Care© Program of the chemical industry associations worldwide? (XO-20) ❏ ❏ ____

  b. Does the facility have an external hazard communication program (i.e., distribution of Material Safety Data Sheets (MSDS's) for products you supply)? (XO-20) If yes, is there a second party (e.g., health professional, legal) review done of your products MSDSs prior to their distribution? (XO-10) ❏ ❏ ____

  c. Does the facility have a system in place for compliance with chemical notification requirements (e.g., TSCA, EINECS/ELINCS, DSL) to appropriate agencies? (XO-20) ❏ ❏ ____

Total for Section II   ____

III. Internal Audits of the Facility Supplying Raychem	Yes	No	Pts*
15. Are formal EHS regulatory compliance audits of the facility's operations conducted? If yes, a. how often? ❏ Every year (30) ❏ Every 2 years (20) ❏ Every 3 years (10) b. done by whom?_____ c. is a written protocol used? (XO-30) If yes, what is the basis for the audit protocol (e.g., Responsible Care, ISO14000)_____	❏ ❏	❏ ❏	____ ____
16. Are formal EHS management system audits of the facility's operations conducted? If yes, a. how often? ❏ Every year (30) ❏ Every 2 years (20) ❏ Every 3 years (10) b. done by whom? _____ c. do you use a written protocol? (XO-30) If so, what is the basis for the audit protocol (e.g., Responsible Care, ISO14000**) _____ ** In draft form as of November 1995.	❏ ❏	❏ ❏	____ ____
Total for Section III			____

IV. EHS Performance and Continuous Improvement at the Facility Supplying Raychem	Yes	No	Pts*
17a What were the worker lost time accident rates for your employees for each of the last 3 years? (e.g., lost time accidents per 100 employees) Year: ___ Rate: ___; Year: ___ Rate: ___; Year: ___ Rate: ___ b. What is the comparable industry worker accident rate? _____			
18. Does the facility evaluate the lost time accident rates of its contractors? (XO-10) Please give an example of last year's accident rate for one of the most frequently used contractors: _____	❏	❏	____
If there is a "yes" answer to questions 19 a–h, please attach a brief description or example.			
19. Within the last 3 years, has the facility:			
a. Had any worker injuries or occupational illnesses requiring a formal onsite investigation by regulatory authorities? (If no, award 20 points)	❏	❏	____
b. Been the subject of any EHS enforcement actions by any government entities, or does the facility have knowledge of any contemplated enforcement actions? If yes, state the results of the enforcement action (e.g., consent order, penalties, no action) and describe the circumstances: ____ _____(If no, award 20 points)	❏	❏	____
c. Been under any regulatory agency orders for corrective actions related to EHS issues or code violations (If no, award 20 points)	❏	❏	____
d. Been required to investigate and/or remediate soil or groundwater contamination? (If no, award 20 points)	❏	❏	____
e. Paid damages as a result of environmental litigation? (If no, award 20 points)	❏	❏	____
Within the last 3 years, has the facility:			
f. Experienced any accidents involving property losses or business interruption losses totaling $100,000 or more? (If no, award 10 points)	❏	❏	____
g. Had an insurer significantly increase premiums or deductible levels because of failure to meet insurance industry standards, or because of high property or business interruption losses? (If no, award 10 points)	❏	❏	____
h. Been the subject of any citizen complaints to regulatory agencies or authorities regarding EHS matters? (If no, award 10 points)	❏	❏	____
i. Received any rewards or commendations for EHS performance? (XO-10)	❏	❏	____

20. Can the facility document continuing EHS performance improvement over the last 3 years in the areas of:			
a. worker protection? (XO-20)	❑	❑	___
b. environmental protection? (XO-20)	❑	❑	___
c. property protection? (XO-20)	❑	❑	___
d. product stewardship? (XO-20)	❑	❑	___
Total for Section IV			___
Grand Total (for Raychem use only)			___

Reprinted with the permission of the Raychem Corp., Menlo Park, CA
Source: Raychem Corporate Environment, Health, and Safety Manual (EHS/3/1128/4.6)

Appendix I
The ISO 14001 Framework

Shows the interrelationship of the various ISO 14001 requirements. The four (4) boxes at the corners represent requirements which are applicable to all of the elements within the flowchart itself (e.g., all are subject to document control, management review, annual EMS auditing and records)

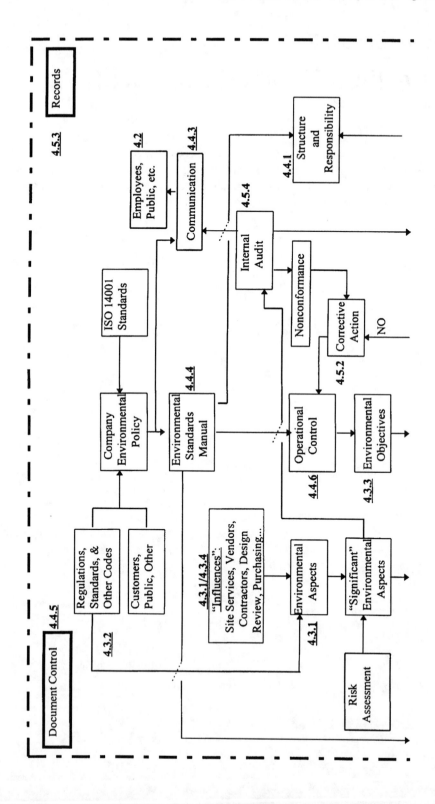

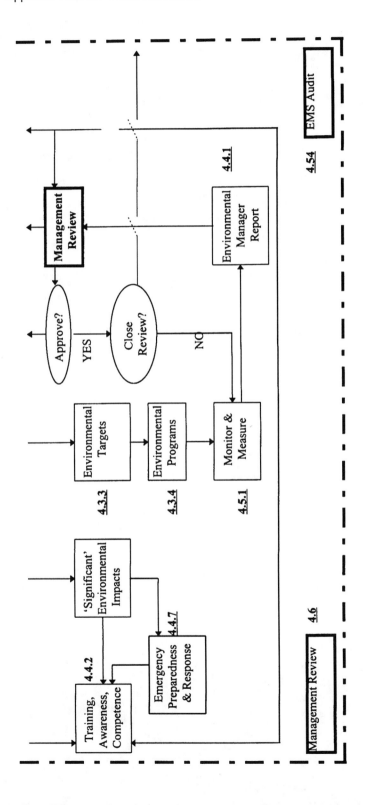

Bibliography

1. ISO 9001: 1994 Copyright® International Standardization Organization
2. ISO 9004: 1994 Copyright® International Standardization Organization
3. ISO 14001: 1995 Copyright® International Standardization Organization
4. ISO 14004: 1995 Copyright® International Standardization Organization
5. QS-9000:1996 Chrysler Corp., Ford Motor Corp., and General Motors Corp.
6. CEEM Information Services, Int. Env. Syst. Update

Index